Neural and Biochemical Networks:
Organization, Development, and Robustness

by

Marcus Kaiser

A thesis submitted in partial fulfillment
of the requirements for the degree of
**Doctor of Philosophy
in Neuroscience**

Approved, Thesis Committee:

───────────────────────────

Prof. Dr. Claus C. Hilgetag

───────────────────────────

Prof. Dr. Herbert Jaeger

───────────────────────────

PD Dr. Rolf Kötter

Date of Defense: July 1, 2005

School of Engineering and Science

Abstract

During the last decades great progress has been made in understanding the structure and function of single nerve cells. However, little is known about the *global* organization of connections between cortical areas and how this structural cortical network relates to function as measured by functional imaging or electro-encephalograms (EEG). In this work, I therefore studied the organization, development and robustness of cortical, as well as other biological networks, using methods of network analysis.

Chapter 1 gives a brief introduction to networks in biology and outlines previous concepts as well as novel approaches of network analysis.

Chapter 2 presents the data sources for biological as well as non-biological networks used in this study. In addition, limitations of the available data and validity of results based on incomplete data sets are discussed.

Chapter 3 examines the organization of neural systems and important constraints for shaping them. The spatial network of macaque cortical connections shows surprisingly many long-distance connections. In addition, area positions in the macaque cortical as well as the *C. elegans* neural networks could be rearranged so that total wiring length can be reduced by up to 64% of the original value. I show that despite their high resource consumption, long-distance connections help to minimize the number of intermediate nodes, which leads to lower time delay, less interference, and higher synchrony in the network.

Chapter 4 deals with the growth of networks in space. I designed a developmental algorithm that takes into account distance between nodes and which yields networks that are alike various spatial networks from metabolic networks to the German highway system. Such spatial growth can generate networks that are comparable to cat and macaque cortical networks both in respect to neighborhood clustering, number of intermediate nodes and total wiring length. In addition, the inclusion of time windows for development can lead to a defined cluster architecture as observed for cortical networks.

Chapter 5 analyzes the measurement of robustness in biological networks as well as possible underlying causes for high robustness towards unspecific removal of edges or nodes. Biological networks are remarkably robust and I tested different measures to predict the effect of removing edges from the networks. In addition, cortical networks show similar behavior after removal of network components as scale-free networks. The existence of clusters and of highly-connected nodes results in enhanced average-case robustness after removal of edges or nodes. Using multiple lesions, I found that in some cases the effect differed from the effect that was predicted from single lesions. Therefore, multiple lesion analysis could become a framework to predict or explain 'unexpected' effects of experimental lesions (or multiple gene knock-outs in metabolic systems).

Chapter 6 presents a general discussion as well as an outlook on future research including possible technical applications of the current results.

As the approach of network analysis is relatively new to the neurosciences, a glossary of key terms is also provided (p. 134).

Acknowledgments

Doing research can be stressful and frustrating, but it can also be rewarding by yielding exciting results. Here I want to thank all the people that gave advice or cheered me up after yet another rejection letter.

Foremost, I want to thank my supervisor Dr. Claus Hilgetag for the freedom and support for my research. It was a pleasure to work with him and he was always available to discuss ideas and results. In addition, he provided me with the necessary skills to survive in academia. Thanks!

Moreover, I want to thank my colleagues Stoyan Kurtev, Nadia Sachs and Roxana Voitcu for helpful discussions and making Neuroworld a lively place. I am also grateful for useful advice from Dr. Herbert Jaeger, Dr. Martin Zacharias, Dr. Alexander Birk, Dr. Holger Kenn, and Mark Schreiber. Also thanks to Erika Schulz who was supporting in every possible way (cookies and chocolate will never be forgotten).

Thanks to Dudi Deutscher for providing me with the MSA Matlab analysis package. Also, thanks to Yoonsuck Choe for contributing spatial positions of neurons in the *C. elegans* neural network. I am thankful for comments and help from Robert Martin, Peter Andras and Malcolm Young who worked together with me on sections 3.2 and 5.3.

I am grateful for financial support from the International University Bremen and the German National Academic Foundation (Studienstiftung) as well as for several conference and workshop fellowships sponsored by various funding organizations.

Finally, I want to thank Hsin-Lin and my parents for support and understanding.

Contents

Abstract i

Acknowledgments iii

1 General introduction **1**

    1.1 Complex systems and networks 1

    1.2 Cortical connectivity networks 3

    1.3 Network analysis and topological features 5

        1.3.1 Small-world networks 6

        1.3.2 Scale-free networks 8

2 Analyzed network data **11**

    2.1 Biological networks 11

        2.1.1 Cortical networks 11

        2.1.2 *C. elegans* neural network 17

        2.1.3 Biochemical networks 18

    2.2 Technical networks 19

        2.2.1 Highway transportation network 19

        2.2.2 Internet (Autonomous Systems level) 20

3 Organization — 22

3.1 Introduction — 22

3.2 Comparison of brain and benchmark network topologies — 25

3.2.1 Introduction — 25
3.2.2 Data — 26
3.2.3 Methods — 30
3.2.4 Results — 30
3.2.5 Discussion — 32

3.3 Long-distance connectivity in the macaque — 34

3.3.1 Introduction — 34
3.3.2 Results — 35
3.3.3 Discussion — 37

3.4 Optimal component placement? — 37

3.4.1 Data — 38
3.4.2 Methods and Results — 38
3.4.3 Discussion — 41

3.5 Alternative wiring constraints — 42

3.5.1 Introduction — 42
3.5.2 Data and Methods — 43
3.5.3 Results — 46
3.5.4 Discussion — 50

3.6 Summary — 53

4 Development   55

   4.1 Introduction 55

       4.1.1 Developmental factors for neural fate 56

       4.1.2 Factors determining neural projection targets 57

       4.1.3 Development of spatial networks 58

   4.2 Analyzed networks as examples for spatial graphs 60

   4.3 Methods for modeling network growth 63

       4.3.1 Distance independent growth models 63

       4.3.2 Distance dependent growth 64

       4.3.3 Spatial growth model 65

   4.4 Results 67

       4.4.1 General properties of spatial growth networks 67

       4.4.2 Distinguishing types of network development 72

       4.4.3 Case study: cortical connectivity networks 75

   4.5 Summary 81

5 Robustness   83

   5.1 Introduction 83

   5.2 Predicting the effect of single edge removal 85

       5.2.1 Introduction 85

       5.2.2 Data and Methods 86

       5.2.3 Results 89

       5.2.4 Discussion 99

   5.3 Effect of sequential node or edge removal 102

	5.3.1	Data and Methods	102
	5.3.2	Results	105
	5.3.3	Discussion	108
5.4	Multiple lesions		111
	5.4.1	Introduction	111
	5.4.2	Data and Methods	112
	5.4.3	Results	115
	5.4.4	Discussion	119
5.5	Summary		121

6 General discussion and outlook — 123

- 6.1 Which constraints shape the brain? … 123
- 6.2 How did the brain develop? … 125
- 6.3 Why is the brain robust? … 128
- 6.4 Possible applications … 133

Glossary — 134

Cortical connectivity matrices — 139

Abbreviations — 142

Bibliography — 149

Publication list — 162

Curriculum vitae — 165

List of Figures

1.1 Examples for cortical networks 5

1.2 Topology of random and scale-free networks 9

3.1 Example of one scale-free benchmark network with 200 nodes that was generated by the modified algorithm. The cumulative degree distribution of the network shows a power-law tail. 29

3.2 Comparison of the degree distribution of cortical and random benchmark networks 32

3.3 Similarity of network connectivity between cortical and scale-free benchmark networks 33

3.4 Approximated fiber length in the macaque cortical network 35

3.5 Wiring organization of the macaque 36

3.6 Component placement optimization of the macaque cortical network 39

3.7 Component placement optimization of the *C. elegans* neural network 41

3.8 Macaque original cortical and alternative minimal wiring 47

3.9 Network properties for different macaque wiring scenarios 48

3.10 Effect of rewiring for macaque cortical network properties 50

4.1　Visualization and degree distribution of the German highway (Autobahn) system 61

4.2　Linear scale-free network generated by spatial growth 68

4.3　Change of small-world properties for different parameter ranges for spatial growth 69

4.4　Parameter space of spatial growth and resulting network classes . . 70

4.5　Spatial growth with additional preferential crowding condition . . . 72

4.6　Network parameter change after spatial growth and growth with preferential attachment 73

4.7　Evolution of the Internet 74

4.8　Time windows and initial seed nodes 78

4.9　Adjacency matrix for spatial growth with time windows 79

4.10　Spatial growth with time windows: degree distribution and 'fiber' length distribution 80

5.1　Edge frequency and resulting damage after edge elimination 92

5.2　Performance of predictors for edge vulnerability in biological, highway and benchmark networks 93

5.3　Edge vulnerability in benchmark multi-cluster networks 95

5.4　Edge vulnerability for different network patterns 97

5.5　Damage after sequential node eliminations in Macaque as well as benchmark networks 105

5.6　Fraction and value of peak ASP for attack node elimination 107

5.7　Fraction and value of peak ASP for attack connection elimination . 109

5.8 Prediction of double lesion effect from single lesions in the macaque cortical network 117

5.9 Multiple-Shapley-Analysis (MSA) for the cat cortical network . . . 118

5.10 Ratio of neutral knockouts for multiple enzyme elimination 119

6.1 Projections between brain regions in the macaque brain network . . 130

A.1 Cat connectivity matrix (55 nodes) 139

A.2 Cat connectivity matrix (65 nodes) 140

A.3 Macaque connectivity matrix (73 nodes) 141

List of Tables

2.1 Overview of different analyzed networks 21

3.1 Comparison of brain networks and benchmark networks. 27

3.2 Properties of modified algorithm for the generation of scale-free networks 29

3.3 Values for ASP, C and total wiring length of cortical and alternative wiring arrangements. 48

4.1 Comparison between cat connectivity network and network generated by spatial growth 76

4.2 Comparison between macaque connectivity network and network generated by spatial growth 76

5.1 Performance of different predictors for edge vulnerability 91

Chapter 1

General introduction

1.1 Complex systems and networks

Of all networks that are large, complex or at least complicated the brain has always been of special interest both to neurobiologists as well as computer scientists because it is the physical correlate of intelligence and consciousness as well as being the most capable parallel processing unit. The brain has evolved to combine sensory information and memories into useful behavioral responses. In addition to making correct choices about what to do next, processing often has to be fast. For example, detecting possibly harmful situations (for example, dangerous animals) and starting a fight or flight reaction needs fast visual object detection and a rapid initiation of movement. In terms of processing, the brain is often compared to computers (von Neumann, 1958). However, whereas standard computer architecture relies on sequential processing steps, the brain is optimized for heavily parallel computation. In this thesis, I will analyze the underlying architecture of the brain at the level of connections between cortical areas. Such cortical networks consist of cortical areas as nodes and fiber connections between them, which can be represented as directed edges. These networks represent a global level of cortical architecture in contrast to local levels such as layers or columns within individual areas or small ensembles of neurons.

Previous approaches to theoretically model biological neural networks used information and control theory. Information theory (Shannon, 1948) was applied to analyze how information about the outside world is encoded in firing patterns of neurons. Observing the firing of multiple neurons simultaneously, for example, can help to determine how much information is encoded by the firing rate of neurons versus how much information is yielded by simultaneous firing (or *not* firing) (Panzeri et al., 1999). Subsequently, using Bayesian statistics, higher cortical areas can infer a presented stimulus from firing patterns of neurons (Andras et al., 2002). At the circuit level, control theory was used for modeling neural systems. Originally developed for industrial applications, control theory was refined under the name cybernetics to describe regulatory circuits in biological systems (Wiener, 1961). Cybernetics, however, only describes a part of the network and neglects the complexity that arises by interacting circuits at the global level. Meanwhile, more information on neural circuitry has become available.

Advances in functional imaging (for example, positron emission tomography/PET, or functional magnetic resonance imaging/fMRI) made it possible to observe activity in the whole brain. Such global information can—at present—hardly be related to local analysis of information processing or circuits. Instead, linking global activity with the global network architecture might be helpful to explain the relationship between structure and function. I will therefore analyze the global level of neural systems, that is, cortical networks in the brain of the cat and the macaque monkey. Using methods applied in other fields such as computer science, sociology or statistical mechanics, I want to determine structural properties of cortical networks and their relation to brain function. The possible function of network elements found by computer simulations can subsequently be validated by experimental studies, for example, lesion studies for cortical networks or gene knock out studies for biochemical networks. However, such experiments are beyond the scope of this thesis. Although I also present results on biochemical networks, the main focus is on cortical networks in the mammalian brain.

1.2 Cortical connectivity networks

Cortical areas are brain modules which are defined by structural (microscopic) architecture. Observing the thickness and cell types of the cortical layers, several cortical areas can be distinguished (Brodmann, 1909). Furthermore, areas also show a functional specialization. Within one area further sub-units (for example, cortical columns) exist, however, these units were not part of the analysis as there is currently not enough information about their connectivity. Using neuroanatomical techniques, it can be tested which areas are connected, that means whether projections in one or both directions between the areas do exist or not. If a fiber projection between two areas was found, the value '1' was entered in the adjacency matrix; the value '0' defines absent connections or cases where the existence of connections was not tested (Fig. 1.1a).

Contrary to popular belief, cortical networks are not completely connected, that means, *not* 'everything is connected to everything else': Only about 30% of all possible connections (arcs) between areas do exist. Instead, highly connected sets of nodes (*clusters*) are found that correspond to functional differentiation of areas. For example, clusters corresponding to visual, auditory, somatosensory and fronto-limbic processing were found in the cat cortical connectivity network (Hilgetag & Kaiser, 2004). Furthermore, 'streams' of visual cortical areas are segregated functionally (Ungerleider & Mischkin, 1982) as well as in terms of their input, output and mutual connections (Young, 1992). Topological sequences of areas indicate potential signaling pathways across the cortical networks (Petroni et al., 2001), and sensory cortical networks may also possess elements of a serial organization (Young et al., 1995). In addition, about 20% of the connections are unidirectional (Felleman & van Essen, 1991), that means, a direct projection from area A to area B but not vice versa exists. Although some of these connections might be bidirectional as the reverse direction was not tested, there were several cases where it was confirmed that projections are unidirectional. Therefore, for calculations I implemented measures that worked for directed graphs (networks including unidirectional links).

I analyzed connectivity of the cat and the macaque (rhesus monkey, Fig. 1.1b) as well as of the small nematode, or roundworm, *C. elegans*[1]. Both networks exhibit clusters, that means, areas belonging to a cluster have many existing connections

---

[1] Other known connectivity data—not used here—is available for the rat. Unfortunately, at present there is not enough information about structural connectivity in the human brain that would allow network analysis.

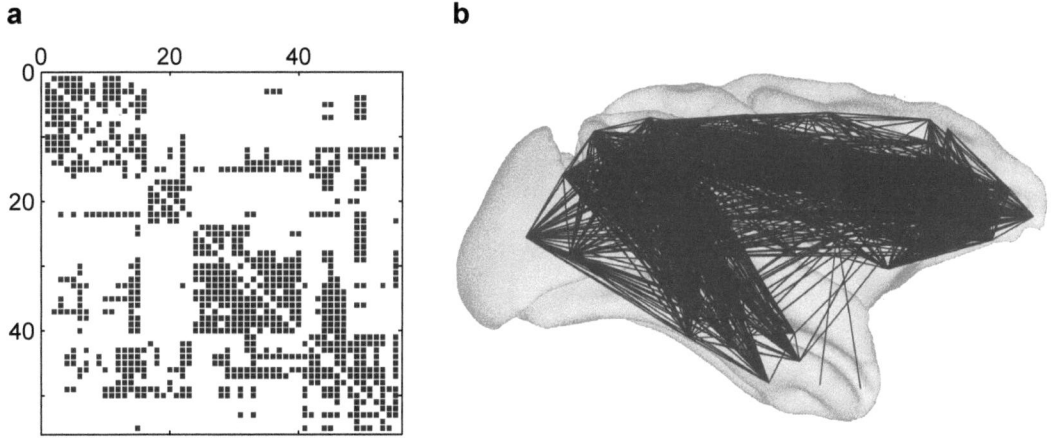

Figure 1.1: **a**, Adjacency Matrix of the cat connectivity network (55 nodes; 891 directed edges). Dots represent 'ones' and white spaces the 'zero' entries of the adjacency matrix. **b**, Lateral view of one hemisphere of the macaque cortex with superimposed cortical connectivity network (95 nodes; 2,402 directed edges).

between them but few connections to areas within different clusters (Young, 1993; Scannell et al., 1995). These clusters are also functional and spatial units. Two connected areas tend to be spatially adjacent on the cortical surface and tend to have a similar function (for example, both taking part in visual processing).

In addition, cortical networks show maximal structural and dynamic complexity which is thought to be necessary for encoding a maximum number of functional states and might arise as a response to rich sensory environments (Sporns et al., 2000b; Seth & Edelman, 2004).

1.3 Network analysis and topological features

Graphs—or networks in colloquial language—consist of objects or *nodes* and relationships or *edges* (Diestel, 1997; Gross & Yellen, 1998). For neural networks, nodes would represent neurons, and edges their connections (that means, synapses). In metabolic networks, nodes could be the substrates or molecules and edges could be reactions between molecules. From several studies during the last few years, one central—and very surprising—result emerged: real-life networks as different as brain connectivity, protein-protein interactions, food webs, social acquaintance networks, the Internet or highway transportation networks appear to share common architectonic principles (Strogatz, 2001; Albert & Barabási, 2002; Newman, 2003). All these networks are sparse, yet possess clusters. Many of them also feature a highly efficient pathway structure, in which only a small number of intermediate connections have to be passed in order to get from one element to another. In relation to social networks this has long been known as the 'small-world' phenomenon where persons are linked with only six degrees (that means, six intermediate persons) of separation (Milgram, 1967). More recently, the growing availability of network data—from topological data about the Internet to the large amount of data on metabolic and regulatory cellular networks—has fueled research under a joint heading of *network analysis* (Albert & Barabási, 2002). Moreover, the field was recently called *network science* to emphasize the interdisciplinary nature of the research. In the following, I will introduce some recent results and concepts of network analysis.

1.3.1 Small-world networks

Many complex networks exhibit properties of small-world networks (Watts & Strogatz, 1998). In these networks neighbors are better connected than in comparable Erdös-Rényi random networks (Erdös & Rényi, 1960) (called random networks throughout the text) whereas the average path length remains as low as in random networks. Formally, the average shortest path (ASP, similar, though not identical, to characteristic path length ℓ; Watts, 1999) of a network with N nodes is the average number of edges that has to be crossed on the shortest path from any one node to another:

$$ASP = \frac{1}{N(N-1)} \sum_{i,j} d(i,j) \quad with\ i \neq j, \tag{1.1}$$

where $d(i,j)$ is the length of the shortest path between nodes i and j. The network is assumed to be a simple graph (cf. glossary), that is, having no loops or multiple edges.

The neighborhood connectivity is usually measured by the clustering coefficient. The clustering coefficient of one node v with k_v neighbors is

$$C_v = \frac{|E(\Gamma_v)|}{\binom{k_v}{2}}, \tag{1.2}$$

where $|E(\Gamma_v)|$ is the number of edges between direct neighbors of v and the binomial coefficient $\binom{k_v}{2}$ is the number of possible edges between all direct neighbors (Watts, 1999). In the following analysis, I use the term clustering coefficient as the average clustering coefficient for all nodes of a network. Note that high clustering coefficient does not automatically coincide with the existence of *multiple* clusters.

Small-world properties were found on different organizational levels of neural networks: from the complete network of the nematode *C. elegans* consisting of 302 neurons (Watts & Strogatz, 1998) to cortical networks of the cat and the macaque (Hilgetag et al., 2000). In addition, social networks, the World Wide Web, and various spatially distributed networks show properties of small-world networks.

1.3.2 Scale-free networks

In addition to small-world features, many real-world networks have properties of scale-free networks (Barabási & Albert, 1999). In such networks, the probability for a node possessing k edges is $P(k) \propto k^{-\gamma}$. Therefore, the degree distribution—where the degree of a node is the number of its connections—follows a power-law. In random networks, degrees only occur at one characteristic scale, that is, possible degrees are near the average number of connections per node. For example, if nodes have 10 connections on average, degrees in the range of 0–30 would occur. For a scale-free network, however, nodes with 1000 or more connections might exist even if the average is still 10 connections. By this, the degree distribution is not limited to one scale but scale-free (Barabási, 2002). In such networks, highly connected nodes or *hubs* can arise that would be highly improbable for random networks (Fig. 1.2). For example, ATP (adenosine triphosphate) would be a hub in metabolic networks as it is a metabolite that takes part in many chemical reactions.

Scale-free networks can be generated by growth and preferential attachment (Barabási & Albert, 1999). Starting with some connected nodes, at each step a new node is added to the network (growth). The new node establishes connections with the already existing nodes. However, nodes with many connections are preferred. This results in an imbalanced degree distribution, in which highly connected nodes of the existing network get even more new connections ("rich get richer"; Barabási, 2002). Moreover, various other methods to yield scale-free networks have been described in the literature (Burda et al., 2001; Caldarelli et al., 2002; Holme & Kimy, 2002; Valverde et al., 2002).

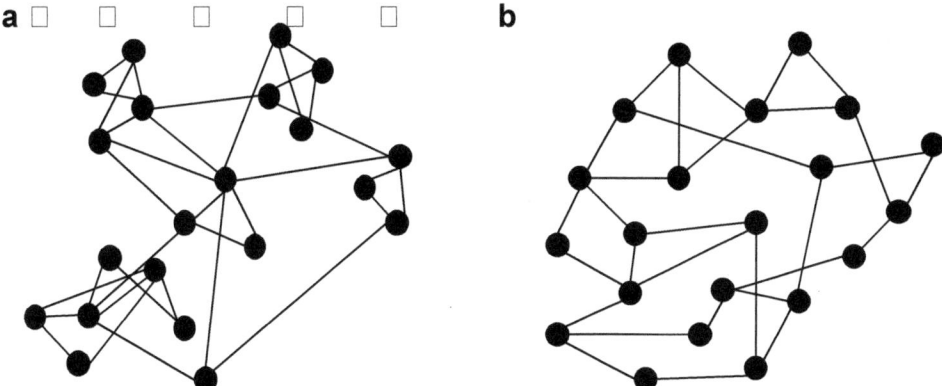

Figure 1.2: Examples of benchmark networks. Schematic view of network connectivity features (both networks have the same number of nodes and edges). **a**, Simple scale-free network having highly-connected nodes (hubs) here shown at the center. **b**, Simple random network with randomly distributed connections and an absence of hubs.

The power-law degree distribution of scale-free networks was found for various real-world systems such as the World-Wide-Web (WWW) (Albert et al., 1999; Huberman & Adamic, 1999), E-Mail networks (Ebel et al., 2002) or the network of human sexual contacts (Liljeros et al., 2001). In addition, biological networks such as the protein-protein interaction network (Jeong et al., 2001; Ho *et al.*, 2002), metabolic networks (Jeong et al., 2000) and brain functional networks (Eguíluz et al., 2005) were scale-free. Hubs can have various functional roles in biological systems. In populations, for example, hubs were found to be selection amplifiers for the evolution of species (Lieberman et al., 2005).

A general question for network analysis is how well the studied systems are represented in the network model and which questions can be answered by this theoretical approach. A network approach might appear as a gross oversimplification of biological systems, because information about the specific characteristics (for example, morphology of individual neurons or stoichiometric rates of chemical reactions) is not represented in the model. However, many important questions about biological systems may be addressed at the network level, for instance, how the connections in a network are related to the function of individual nodes or the whole network, what the rules are for establishing a connection between two elements, or which elements of a network are most important for network performance. Previous studies have shown, for example, that it is possible to identify essential reaction paths based on information about metabolic networks (Schuster & Hilgetag, 1994; Schuster et al., 2002; Stelling et al., 2002). In the next chapter, I will present the networks that were used in this study as well as questions related to the quality of data acquisition.

Chapter 2

Analyzed network data

In my studies, I analyzed several biological and artificial networks to identify which properties are common for all networks and which are specific only for cortical or biological networks. As each data set was used in several sections, the descriptions are given here and subsequent method sections only refer to this chapter. Table 2.1 at the end of this chapter gives an overview of network properties.

2.1 Biological networks

2.1.1 Cortical networks

In cortical networks, nodes are brain regions or areas of the cerebral cortex, such as primary visual cortex V1, defined by their morphology and cytoarchitectonics, and the connections are the long-range (for example, corticocortical) fibers that connect two regions, as found by neuroanatomical tracing studies.

Identifying cortical areas and connections between them

All these studies rely on the concept of cortical areas. Areas are anatomical units of the cortex which can be distinguished by anatomical features of the cytoarchitecture. Looking at the cellular topology, areas in the human cortex could be distinguished by different layer patterns (for example, the primary visual cortex/area 17 exhibits a distinct stripe—the Gennari stripe—and is thus called striate visual cortex) or by different cell types (for example, giant Betz pyramidal cells only occurring in the primary motor cortex/area 4) (Zilles & Rehkämper, 1998). An early parcellation of human cortical areas was achieved by Brodmann (1909) who distinguished 47 cortical areas. Such methods have also been used to identify areas in the macaque and the cat.

The existence of connections between areas was verified by tract-tracing or neuronographic studies. During tracing studies, a dye is injected into one area *in vivo* and is transmitted along the axons of neurons. Depending on the used substance, the transport can occur anterogradely (from the soma or cell body of the neuron to the synaptic end plates), retrogradely (taken up at the synapses and projected back to the soma), or bidirectionally. In subsequent anatomic analysis of cortical areas, not only the area where the injection took place will show the dye, but also areas which receive projections from that area or that project to that area. One problem of this method is that it is difficult to decide whether connections exist between nearby cortical areas as the dye could reach the neighbor not only through the axonal projections but also simply by diffusion through the surrounding medium. In addition, long-distance connections might be missed if only few nerve fibers run between the two areas. Neuronographic mapping, as a second method, is semi-functional in that it observes activity spreading along cortical areas. Applying strychnine to one area decreases local inhibition and thereby induces epileptiform activity. Steady-state epileptiform activity is first observed in the concerned area and subsequently in distant areas conveyed by corticocortical projection fibers. Neuronographic studies for the macaque monkey revealed similar network clusters as the analysis of connectivity networks based on the tract tracing method (Stephan et al., 2000).

Recently, another method—Diffusion Tensor Imaging (DTI) (Tuch et al., 2003)—was proposed for the study of human cortical connectivity. DTI has a huge advantage in that it is, in contrast to tract tracing, non-invasive and does not need a *post-mortem* analysis.

Validity of cortical connectivity data

In this thesis, I use connectivity data based on tract tracing studies. Whereas the problems of the method for identifying connections were discussed in the previous section, it remains to be seen to what extent the data set of cortical connectivity is complete and whether missing connections will change the conclusions drawn from the present analysis. For example, there have been more attempts to find connections in the cat brain compared to the macaque. This is indeed one explanation why the network density—the percentage of existing compared to all possible connections—is almost twice as high in the cat (30%) compared to the macaque (15%). Within species, some subsystems of the brain have been more thoroughly studied than others. The connectivity of the macaque motor or visual system received more attention from anatomists than other parts of the brain. For example, in neuronographic studies the range of experiments of areas that were most and least examined approximated a factor of seven (Stephan et al., 2000). Another problem is that most studies in the macaque only investigate ipsilateral connections, therefore, the datasets mostly comprise the internal connectivity of one hemisphere rather than that of the complete brain (Kötter, 2004).

In addition to connections that were not tested for, some potential connections were explicitly reported absent. In my analysis, I made no difference between such absent connections and connections that were not tested (both were assigned the value '0' in the adjacency matrix). Not only is this consistent with the treatment used in other studies (Scannell et al., 1995; Sporns & Kötter, 2004), but also justified in that different strategies for handling absent connections led to very similar results (Young, 1992; Young et al., 1995).

In order to avoid misleading conclusions, which could result from untested or non-detected missing connections in the network, two strategies were applied. First, both for the cat and the macaque, the whole network was used for analysis. Therefore, no result is based on potentially less well-characterized subsystems. As the well-characterized visual system is always included in the network, the effects of other subsystems on the analysis should be lower compared to a scenario where individual subsystems are evaluated. Still, connections missing in the data set could—when included—change the cluster architecture either by being part of a cluster and therefore supporting the cluster partition, or by lying between clusters and therefore eroding the current separation into different clusters. Therefore, as a second strategy, the results for the less completely tested macaque connectivity were compared to the well-characterized cat connectivity whenever possible[1].

Cat and macaque cortical connectivity networks

I used cat and macaque cortical connectivity data (Scannell et al., 1995; Scannell et al., 1999; Young, 1993; Kötter, 2004). In both species, the data comprised connections between cortical regions and some sub-cortical structures (for example, the amygdala).

For the macaque brain, I considered 73 brain structures with 837 connections between them (Young, 1993). In the case of the cat brain, I considered 65 structures and 1,139 connections (Scannell et al., 1995). The edge density of the macaque brain graph, that is, the number of reported connections divided by the number of all possible connections, is 16%. For the cat brain with 65 areas, the edge density is 27%. There are on average 22.8 connections for each structure in the macaque brain and 35 connections for the cat brain structures. In addition, for some analysis a smaller cat connectivity data set with 55 areas was used (Scannell et al., 1999).

The adjacency matrices of the cat networks (55 or 65 nodes) and the macaque network (73 nodes) are provided in the appendix on page 139.

---

[1] A comparison was impossible for the analysis of spatial positions of cortical areas, as only positions for the macaque brain could be computed using the Caret software. At the moment, the Caret software does not include surface coordinates of the cat brain.

Spatial area positions in the macaque

In addition to the connectivity of the macaque, also spatial positions of cortical areas could be included in the analysis. In that data set, 95 cortical areas and subareas of the macaque monkey with corresponding positions were available. The connectivity data were based on three compilations of neuroanatomical tracer studies (Carmichael & Price, 1994; Lewis & Van Essen, 2000; Felleman & van Essen, 1991) available at http://www.cocomac.org (Kötter, 2004). Average spatial positions of cortical areas were estimated based on surface coloring using the CARET software (van Essen Lab, http://brainmap.wustl.edu/caret). Areas for which no spatial position in the CARET software package existed, or which were duplicate in the joint area set, were removed. The length of connections was approximated as the direct Euclidean distance between the geometric centers of two connected areas. Naturally, this only represents an average case and individual fiber tracts might be longer or shorter for real cortical connections. Fibers between direct neighbors on the cortical surface might be shorter than is suggested by the distance of their centers. On the other hand, cortical folding, that is the pattern of gyri and sulci, might make a direct fiber connection impossible, leading to curved fiber trajectories and therefore longer fiber length than assumed. In addition, brain folding *after* fiber establishment as well as ventricles or other fiber bundles which hinder a straight projection can lead to longer pathways. Nonetheless, the approach can lead to a first approximation of wiring organization at the global level as the findings did not rely on the length of individual projection fibers. Rather, the number of long-range connections together with the approximated total length

CHAPTER 2. ANALYZED NETWORK DATA

of cortical connectivity was considered important.

2.1.2 *C. elegans* neural network

I also analyzed the neural network of the nematode (roundworm) *C. elegans* as an example for smaller networks (note, that in contrast to cortical networks as part of the central nervous system, the *C. elegans* network encompasses the complete neural system of that organism). The number of somatic cells of the nematode was 959, including 302 neurons. Fortunately, the synaptic connectivity of all 302 neurons of the nematode had been previously described (White et al., 1986; Durbin, 1987) making network analysis of the whole neural system possible.

Furthermore, Dr. Yoonsuck Choe (Texas A& M University, USA) kindly provided the spatial positions of all neurons enabling approximations of wiring lengths (Choe et al., 2004). Spatial positions were estimated from the drawings of neuron positions in White et al., 1986. The position data was two-dimensional along the lateral plane. The potential influence of the missing third axis on the approximated axon length is low, as the diameter of the worm is much smaller than its length ($80\mu m$ compared to $1300\mu m$; White et al., 1986). Naturally, this does not prevent that large errors could have occurred for the few local connections where the elongation along the longitudinal plane is lower than along the transversal plane check planes. However, as for the cortical networks, the analysis mainly relied on the proportion of long-distance connections and the approximated total wiring length of the system.

For my analysis, I only included chemical synapses as connection of neurons disregarding electrical synapses between neighboring neurons. Neurons with zero degree—that means, no chemical synapses—were removed from the analyzed network. Therefore, only 256 out of 302 neurons were part of the final data set.

2.1.3 Biochemical networks

In the case of biochemical network studies, various extensive datasets about cellular metabolic pathways as well as protein-protein interaction networks have become available lately and some were used for analysis. For *metabolic networks*, molecules are the nodes and reactions are the edges of the network. A data set about metabolic networks in 43 organisms (Ravasz et al., 2002) was used. The data sets are based on data deposited in the WIT database (`http://igweb.integratedgenomics.com/`). In the database, the existence of a metabolic pathway was predicted from the annotated genome of an organism in combination with established data from the literature. As not all genomes were fully sequenced, the number of substrates and reactions will be lower in the data set than in the actual metabolism of many organisms (Jeong et al., 2000). However, the data sets already provide sufficient data for a first analysis.

The yeast *protein-protein interaction network* consists of proteins and interactions between them found by Two-Hybrid tandem mass spectrometry (Uetz et al., 2000; Schwikowski et al., 2000; Ito et al., 2001). The protein-protein interaction network consisted of 149 disconnected components. The main component contained 79% of the proteins, and the remaining components mostly were composed of only one pair of proteins. Only the largest connected component was analyzed to avoid unreachable pairs of nodes and subsequent problems with path length calculations. The analyzed component consisted of 1,846 proteins and 4,406 interactions. With a value of 6.8%, the clustering coefficient was considerably smaller than for brain networks. As shown before (Jeong et al., 2001), connections in this network are also distributed in a scale-free fashion.

However, one has to remember that the yeast two-hybrid method yielding protein-protein interaction data (Schwikowski et al., 2000; Ito et al., 2001; Gavin *et al.*, 2002) produces many artifacts (Kitano, 2002) which might have influenced the results. Recently, several methods have been discussed to identify or predict true protein-protein interactions (Goldberg & Roth, 2003; Shon et al., 2003). Because establishing veridical large-scale interaction data is still work in progress, I have—as in other studies concerning network analysis (Jeong et al., 2001; Han et al., 2004)—used the raw data yielded from two-hybrid mass spectrometry. Despite potential problems with the data set, my results from the protein-protein interaction network were consistent with results from the better characterized metabolic networks.

2.2 Technical networks

2.2.1 Highway transportation network

As an example for a man-made transportation networks, the German highway (Autobahn) system was explored. The data of location nodes and connections was processed from the 'Autobahn Informations System' (AIS), as accessible under `http://www.bast.de` (data as of 18 July 2002). Only roads defined as highways were included in the analysis. Multiple highway exits for the same city (for example, Hagen-West and Hagen-Nord) were merged to one location representing the whole city as a node of the network. Parking and resting locations were excluded from the set of nodes. Due to the merging process and highways currently under construction, eight percent of the nodes were separated from the largest cluster and were excluded from analysis. The resulting network consisted of 1,168 highway exits or locations and 1,243 highway connections (undirected edges) between them.

2.2.2 Internet (Autonomous Systems level)

Another example for technical networks, now transporting information, is the Internet. Whereas both the brain and the Internet transmit information, one has to remember the fundamental differences in network function. Internet routers can choose one out of several alternative pathways for delivering data packages. In the brain, pathways are less flexible (except after gross reorganization after lesions) in that nodes are specialized for a certain function: visual information, on their way to visual areas, will not be transmitted over nodes that deal with auditory information. As intermediate nodes with different function are not an option, more direct connections are needed. This explains why the Internet can work with relatively few connections (see below) whereas the brain is densely connected.

The Internet can be analyzed both on the router and the autonomous systems (AS) level. Whereas routers are the basic units redirecting packages, AS's as independent subnetworks can be identified at a global scale (Tanenbaum, 2003). For example, the DFN ("Deutsches Forschungsnetzwerk", German research network),

Table 2.1: Overview of different analyzed networks with number of nodes N and edges E (number of directed edges for directed networks), density d, clustering coefficient C, average shortest path ASP, maximum node degree (counting the number of directed edges for directed networks), and the maximum length of the shortest path between two nodes. Empty fields denote values that were not calculated.

	N	E	d [%]	C [%]	ASP	max. deg.	max. length
Biological							
Cat_{55}*	55	891	30	55	1.8	62	4
Cat_{65}*	65	1,139	27	54	1.9	71	4
Macaque_{73}*	73	837	16	46	2.2	78	5
Macaque_{95}*	95	2,402	27	64	1.9	111	5
Yeast (PPI[1])	1,846	4,406	0.13	6.8	6.8	18	19
Metabolic[2]	422	1,972	1.3	16	—	194	—
Technical							
Autobahn	1,168	1,243	0.18	0.12	19.4	12	57
Internet	6,524	13,984	0.07	67	3.6	1,512	9

[1] Protein-protein interaction network (largest component).

[2] Average values for 43 organisms.

* Directed network in that E refers to the number of directed edges and maximum degree to the maximum number of directed edges of a node.

through which the International University Bremen is connected to the Internet, would be one node in the AS network. In this study, I analyzed a reconstruction of the global AS network based on snapshots from December 1997, December 1998, and December 1999 (Data from the Measurement and Network Analysis Group; http://moat.nlanr.net/AS/). The data consisted of 6,524 AS and 13,984 (bidirectional) edges between them. The connection density was 0.07% with a higher neighborhood connectivity (clustering coefficient 67%).

Chapter 3

Organization

3.1 Introduction

Global connectivity

Investigations of the global structural organization of neural systems connectivity are a fundamental starting point for understanding structure-function relationships in the brain.

Cortical connectivity may be characterized by various structural network indices, such as symmetry, which describes the proportion of reciprocal connections of an area, and transmission, that is, the ratio of the number of local inputs to outputs (Kötter & Stephan, 2003). Also, a matching index can be computed that assesses the pairwise similarity of areas in terms of their specific afferents and efferents (Hilgetag et al., 2002). These indices are complemented by functional (entropic) measures such as 'segregation' and 'integration' that can be employed to evaluate distributed system performance (Sporns et al., 2000a).

There are several functional and structural features of axonal projections that can be analyzed. On a functional level, it can be observed whether projection synapses release inhibitory or excitatory neurotransmitters. For global connectivity it can be determined whether two connected areas have a similar function or at least process the same modality (for example, visual but not auditory or somato-sensory information). Looking at the structural characteristics, the length of projections or—for corticocortical connectivity—the thickness of fiber bundles, indicating the number of axonal projections, can be determined.

Connection length organization

One structural feature of neuronal networks is that projections tend to prefer nearby neurons rather than distant ones (the higher likelihood for short-distance connections will be of importance in the following chapter on network development). For example, in the rat visual cortex, the probability to establish a connection with another neuron decreases with the distance between the two neurons (Hellwig, 2000). This was also the case for the distribution of axonal lengths of pyramidal cells in the mouse cortex (Braitenberg & Schüz, 1998, Fig. 34). In contrast, a recent study on neuronal connectivity between neurons in layer 5 of the rat visual cortex yielded a low distance-dependence (Song et al., 2005). Future studies need to clarify whether distance-dependence is limited to certain layers or cell types, or is a universal wiring principle (see also Binzegger et al., 2004).

It was argued that not only are individual connections preferably short—mostly connecting to neighboring neurons or cortical areas—but that the network of connections is the shortest possible one (Cherniak, 1994). With such component placement optimization, neurons (or cortical areas) are arranged in a way that every exchange of neuron (or area) positions, while the connections of each node remain the same, leads to an increase of the total connection length as being the sum of all individual connection lengths. Using non-metric approximations for the distance, this was shown to be true for the arrangement of ganglia within *C. elegans* as well as for a sub-set of prefrontal cortical areas (Cherniak, 1994; Klyachko & Stevens, 2003). Saving wire is important in terms of reducing resource consumption due to the establishment of long-distance connections and the energy-consuming propagation of action potential along these connections.

Before exploring the wiring organization, I will first examine the network topology of cortical networks in comparison with different benchmark networks in section 3.2. I will show that cortical networks share several properties with scale-free benchmarks networks even though a direct comparison of the degree distributions was difficult. Then, I will analyze the wiring length pattern of the spatial cortical network of the macaque monkey. In section 3.3, I will observe the distribution of fiber lengths on the global scale showing a quite large number of long-distance projections. In section 3.4, I will show that optimal component placement does *not* occur when the proper metric distance is taken into account for the macaque cortical network of connections between cortical areas as well as for the network of *C. elegans* consisting of connections between individual neurons. Thus, the current dogma about the wiring organization of neural networks appears to be wrong. In section 3.5, I will look at alternative constraints that shape network topology.

Specifically, I will show that the network is optimized for low average shortest path—indicating the number of intermediate areas—in addition to saving wire length. This additional constraint enhances synchronous processing (and thereby the possibility of feature binding), reduces time-delays, and improves the signal to noise ratio.

3.2 Comparison of brain and benchmark network topologies

3.2.1 Introduction

Small-world and scale-free networks are important network classes in the description of real world phenomena. Small-world networks consist of well-connected local neighborhoods with fewer long-range connections between neighborhoods. It has been shown before that the neural network of *C. elegans* (Watts & Strogatz, 1998) as well as cortical networks of the cat and the macaque (Hilgetag et al., 2000) show properties of small-world networks. Scale-free networks, on the other hand, are characterized by their specific distribution of connectivities. The distribution follows a power law, with generally a higher probability for well-connected nodes than in random networks. The small-world and scale-free properties are compatible, but not equivalent (Amaral et al., 2000). Here, I attempted to find out whether cortical networks are similar to scale-free networks. This is complicated by the fact that cortical networks have few nodes and thus a direct observation of a power-law in the degree distribution is impossible.

3.2.2 Data

Structural brain connectivity data

I used cortical connectivity networks of the cat with 65 areas and the macaque with 73 areas (cf. section 2.1.1 for details).

Benchmark networks for comparison

I constructed rewired, scale-free, random, and small-world networks to match the statistical properties (number of nodes and edges as well as similar clustering coefficient and ASP where possible) of the corresponding two brain structure networks (Tab. 3.1).

Rewired networks were derived from the original cortical networks of the cat and macaque by randomly exchanging connections such that the total number of connections of each node remained the same. That means that the rewired networks exhibited the same degree distributions as the original cortical networks.

The algorithm to generate scale-free benchmark networks was based on Barabási and Albert (1999). However, in a modification of their approach I began with an initial graph of six and eight fully connected nodes respectively for the macaque and cat benchmark networks. This was necessary in order to ensure that the clustering coefficient of the initial graph matched the highest clustering coefficient found in the corresponding brain network. As proposed by Barabási and Albert, further nodes were added one by one to the graph by preferential attachment. At the beginning of this process, the probability that a new node is connected to an

Table 3.1: Comparison of brain networks and benchmark networks. The table shows the average shortest path and the clustering coefficient statistics for the macaque and cat brain structure networks, and for the respective benchmark random, rewired, small-world, and scale-free networks. For the benchmark networks, the data shows the mean value and the standard deviation of 50 generated networks.

	Average shortest path	Clustering coefficient
Macaque	2.2110	0.4558
Random mean	1.9940 ± 0.0067	0.1547 ± 0.0032
Rewired mean	2.0469 ± 0.0082	0.2953 ± 0.0108
Small-world mean	2.3748 ± 0.0387	0.4647 ± 0.0174
Scale-free mean	2.0062 ± 0.0259	0.4351 ± 0.0494
Cat	1.847	0.5415
Random mean	1.7440 ± 0.0018	0.2636 ± 0.0032
Rewired mean	1.7969 ± 0.0062	0.3939 ± 0.0059
Small-world mean	1.9026 ± 0.0188	0.5023 ± 0.0113
Scale-free mean	1.7627 ± 0.0131	0.5286 ± 0.0526

existing node i is

$$P(i) = \frac{k_i}{\sum_j k_j}$$

where k_j is the number of connections of the node j. After establishing a connection to node i^*, I recalculated the probabilities to reflect the nature of the scale-free networks: if i is connected to j, then it is more likely that i is connected to nodes which are already connected to j and it is less likely that i is connected to nodes which are not connected to j. The rescaling was undertaken according to

$$P^*(i) = \begin{cases} k_{i^*} P(i) & \text{if nodes } i \text{ and } i^* \text{ are connected} \\ P(i) & \text{otherwise} \end{cases}$$

$$P(i) = \frac{P^*(i)}{\sum_j P^*(j)}$$

The probability for the connections in both directions is the same.

Did this modified procedure still generate scale-free networks? As this was difficult to answer for small networks, I generated larger networks (100, 200, 400, and 800 nodes) where the number of directed connections remained the same (1,485 arcs). For each network size, 50 networks were generated. A power-law behavior was visible for the cumulative degree distribution (Fig. 3.1). The power-law coefficient γ of the cumulative degree distribution was measured as well as the correlation r between degree distribution and power-law fit (Tab. 3.2). As can be seen, the modifications did not change the scale-free nature of the generated networks.

For standard Erdös-Renyi random networks (Erdös & Rényi, 1960)—called random networks in the remaining section—, I began with a fixed number of nodes

Table 3.2: For each network size 50 scale-free benchmark networks were generated. The mean and standard deviation of γ, r, and the clustering coefficient are shown. The edge density of all networks was the same for each network size, therefore, no standard deviation is given.

Nodes	100	200	400	800
γ	0.97 ± 0.07	0.99 ± 0.06	1.02 ± 0.06	1.10 ± 0.06
r	0.88 ± 0.01	0.95 ± 0.02	0.97 ± 0.02	0.96 ± 0.02
Clustering coefficient	0.35 ± 0.06	0.29 ± 0.02	0.17 ± 0.02	0.11 ± 0.01
Edge density	0.15	0.04	0.01	0.00

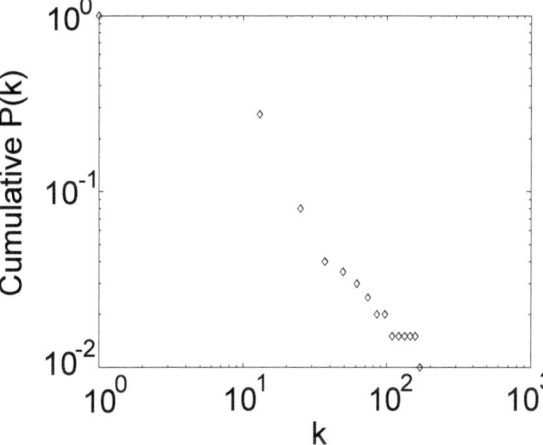

Figure 3.1: Example of one scale-free benchmark network with 200 nodes that was generated by the modified algorithm. The cumulative degree distribution of the network shows a power-law tail.

and randomly generated connections between them. The connection density in the corresponding brain networks, 16% of the number of all possible connections for the macaque and 27% for the cat, governed the presence or absence of any connection in the random benchmark networks. Thus, the degree distribution followed a binomial probability distribution.

Small-world networks were generated by rewiring regular networks as described in the literature (Watts & Strogatz, 1998). The rewiring probability was adjusted so that the resulting networks had similar clustering coefficients as the respective cortical networks (Tab. 3.1).

3.2.3 Methods

Graph similarity

First, nodes were permuted according to their degree distribution (sorted by number of connections of a node). Second, permutated cortical and benchmark matrices were compared by testing what percentage of directed edges in the adjacency matrix occurred at the same position in both matrices. This percentage was defined as the graph similarity S between graph A and B given the number of (directed) connections $|E|$:

$$S = \frac{\sum A \wedge B}{|E|}$$

where $\wedge$ was an element-by-element multiplication with an element in the resulting matrix non-zero if both elements were non-zero; $\sum$ was the sum of all elements in the matrix and thus yielded the number of directed edges existing in both matrices, as these were denoted by a value of one in the matrix.

3.2.4 Results

Degree distribution of cortical networks

Fig. 3.2 shows the degree distributions of macaque and cat compared with random connectivity. In comparison to the random network, the macaque cortical network had highly connected nodes but also more sparsely connected nodes. This is also true for the cat degree distribution which showed a remarkable number of areas with few connections compared to a randomly wired network.

However, as there were only 73 nodes for the macaque and even fewer (65 nodes) for the cat cortical connectivity, the degree distribution only encompassed two orders of magnitude. Therefore, the degree distribution cannot determine whether the network was scale-free or not and I had to use indirect measures to determine whether cortical networks are similar to scale-free or other benchmark networks.

Graph similarity

One possibility was to compare the adjacency matrix of cortical networks with different benchmark networks (Fig. 3.3). For rewired cortical networks, the percentage of identical edges was 28% for rewired macaque and 39% for the rewired cat network. One possible reason for the higher similarity of the rewired cat network was its higher edge density. With more edges in the two compared networks, the probability to have a similar connection also increases leading to a higher similarity. Interestingly, benchmark scale-free networks were as similar to the cortical networks as the rewired cortical networks. In contrast, the similarity of random and small-world networks was significantly lower. This could be attributed to the degree distribution of scale-free and cortical networks being comparable as the rewired network only had the degree distribution in common with the original cortical network.

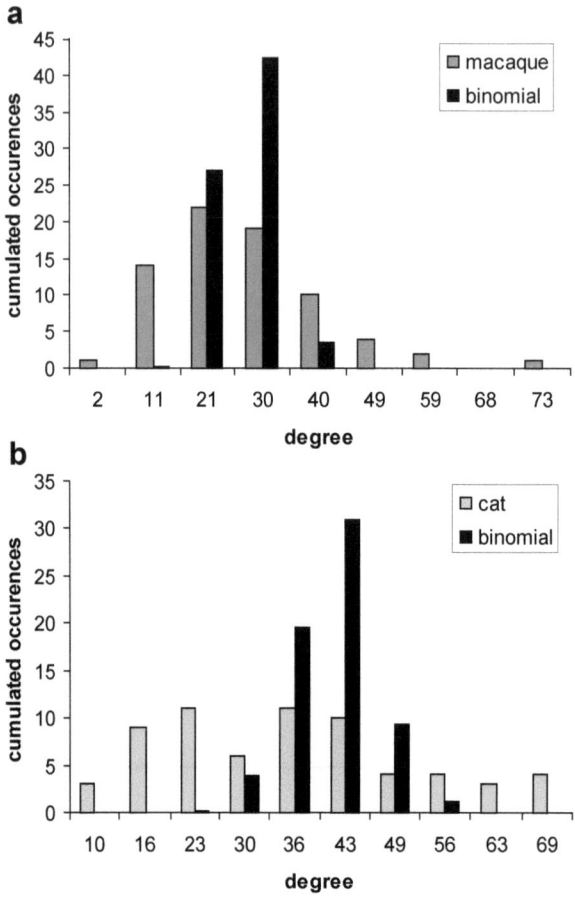

Figure 3.2: Direct comparison of degree distributions of cortical and random networks. **a**, Histogram of the degree distribution of the macaque (gray) compared to the degree distribution of a same-size random network (binomial distribution given the probability p=0.15 that an edge occurs, black). **b**, Histogram of the degree distribution of the cat (gray) compared to the degree distribution of a same-size random network (binomial distribution given the probability p=0.3 that an edge occurs, black).

3.2.5 Discussion

I have compared brain inter-area connectivity networks with different types of benchmark networks, including random, scale-free, and small-world networks, and found indications that the brain connectivity networks share some of their structural properties with scale-free networks.

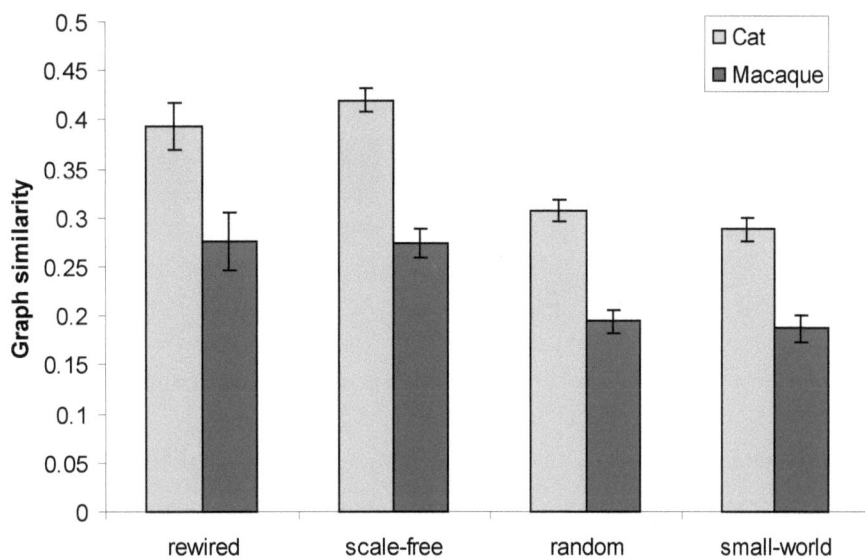

Figure 3.3: Similarity of network connectivity between cortical and benchmark networks. For each type of benchmark network, 1,000 networks were generated. As the cat network has a higher number of edges, the percentages of similar edges are also higher. Interestingly, the similarity with the cortical networks is as good for the scale-free networks as for the rewired cortical networks. In contrast, the similarity of random and small-world networks is significantly lower.

One important feature of this approach is that the rigorous checking of a series of benchmark networks allows assessing the significance of any similarities to other network types found. In the study of a much simpler brain network, it has previously been established that the brain of *C. elegans* is small-world, but not scale-free (Amaral et al., 2000). Furthermore, the similarity measure between scale-free and original cortical networks was higher than for other benchmark networks. Indeed, a scale-free network architecture has been found for functional brain networks in humans (Eguíluz et al., 2005), which might now be explained by underlying scale-free structural connectivity.

Still, the classical evaluation by observing a power-law degree distribution was impossible for cortical networks. In addition to the analysis described here, the effect of the removal of nodes or edges for cortical networks was most similar to that for scale-free benchmark networks (shown later in section 5.3). Together, this shows that cortical and scale-free share similar features such as the occurrence of highly connected nodes. Future research that uses more nodes by either including more cortical (sub-)regions or different levels of organization (e.g., including columns within one area as network nodes) would help to verify a scale-free architecture.

3.3 Long-distance connectivity in the macaque

3.3.1 Introduction

Whereas previous studies concerning metric fiber lengths only looked at the local projection pattern *within* a cortical area, I analyzed for the first time[1] the metric projection length pattern of the global connectivity network between brain areas. I analyzed the connectivity between cortical areas and subareas of the macaque monkey (see section 2.1.1). The resulting network consisted of 95 cortical regions and 2,402 connections (Fig. 3.4a).

---

[1] There were previous studies on connectivity between brain areas (Cherniak, 1994; Klyachko & Stevens, 2003). However, these studies were using neighborhood relations or other indirect measures instead of metric Euclidean distance and were limited to a small subset of cortical areas.

3.3.2 Results

The distribution of distances between nodes, that is, the approximated length of connections showed that also long-range connections occurred (Fig. 3.4b). Long-distance connections existed, for example, between areas 10o and 7a, V2 and 46, and V3 and 46. These connections give a cautioning to predictors of connectivity that are solely based on close proximity of cortical areas (Young, 1992) or criteria of optimal wiring (Klyachko & Stevens, 2003).

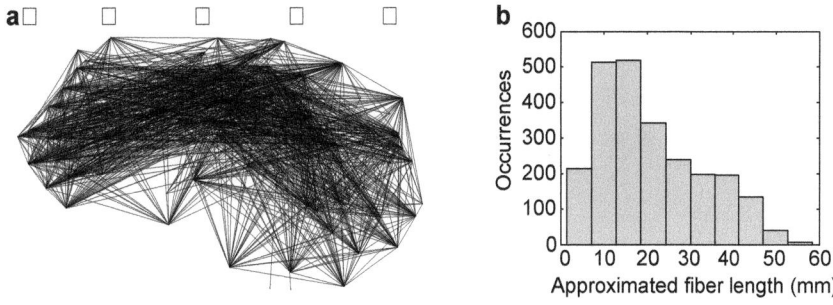

Figure 3.4: **a**, Analyzed macaque cortical network with 95 nodes and 2,402 connections. **b**, Distribution of fiber lengths approximated by the direct distances between connected areas.

A component placement approach has been used previously to demonstrate optimal wiring for *C. elegans* and cerebral cortex (Cherniak, 1994; Chklovskii et al., 2002; Klyachko & Stevens, 2003). In the component placement concept, connections remain fixed and the positions of nodes are evolved. I used a complementary method, in which the positions of areas were invariant, and their connections could be rearranged.

For the macaque cortical network (Fig. 3.5a), the total length of all connections is shown as a bar. The horizontal marks represent the range of possible wiring lengths. The lowest mark in the diagram represents minimal wiring, in which nodes

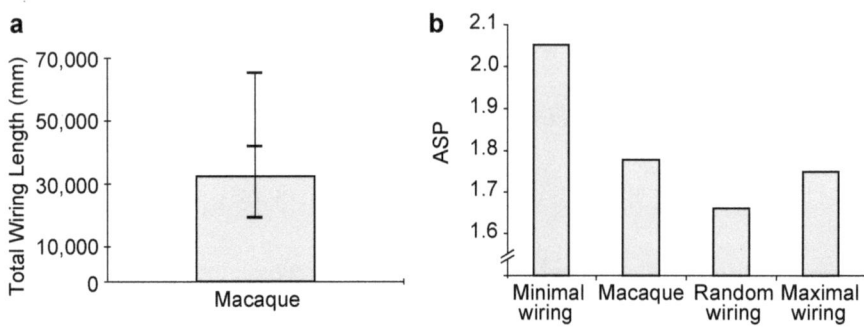

Figure 3.5: **a**, Total wiring length in cortical network. **b**, ASP in the cortical network (Macaque) as well as for alternative wiring configurations with same spatial area positions but different connectivity.

were linked in the order of the shortest distances between them. The upper mark, on the other hand, stands for maximal wiring, in which nodes were connected via the largest-possible distances. The average value of random wiring is indicated by the middle mark. In this case, the distance between nodes was not taken into account for their wiring, resulting in an intermediate outcome between minimal and maximal wiring. For the original cortical network, the total length was positioned between minimal and random wiring, therefore indicating a general preference for short-distance connections.

In addition, I compared a central topological network measure, the average shortest path (ASP), for different metric wiring configurations (Fig. 3.5b). The ASP for minimal wiring was larger than for the actual networks. A low ASP, that is, a low number of intermediate nodes in cortical paths, might be important for several reasons: first, intermediate nodes carry out additional signal transformations and may introduce noise and, second, additional areas and their intrinsic synapses add to the signaling delay in the path. Moreover, without long-range connections, synchronous processing in proximal and distant areas might be impossible. Therefore, preserving a low ASP may be an important constraint for the evolution of cortical wiring. This might explain why long-distance connections occur and lead to a non-minimal wiring organization.

3.3.3 Discussion

In a complementary approach to component placement optimization, I found that the total wiring length characteristics are between random and minimal organization. The existence of long-range connections in these networks increased total wiring length, but also led to a shorter ASP. As the ASP in the extended macaque dataset as well as for the cat data set was at a low value of 1.7, the conservation of a low ASP, that is, a low number of intermediate nodes in cortical pathways, might be a more important constraint than minimizing total 'cable' length.

3.4 Optimal component placement?

Neural networks, such as ganglia in *C. elegans* as well as cortical networks in mammalian brains, have been reported to show optimal component placement (Cherniak, 1994; Chklovskii et al., 2002; Klyachko & Stevens, 2003). This means that changing the position of neurons or cortical areas while keeping the connectivity patterns unchanged, leads to an increase of total wiring length in the network. Evolutionary optimization was discussed as a mechanism for achieving such optimal node positions (Cherniak et al., 1999). In the following analysis, I investigated the wiring organization of neural networks in metric space, analyzing three-dimensional positions of cortical areas in the macaque and two-dimensional projections of positions of individual neurons in *C. elegans*. Interestingly, at both levels of organization no optimal component placement was found.

3.4.1 Data

I analyzed the connectivity between 95 cortical areas and sub-areas of the macaque monkey brain as introduced in the previous section (cf. section 2.1.1). In addition, spatial two-dimensional positions (in the longitudinal plane) of individual neurons in *C. elegans* (White et al., 1986; Durbin, 1987) were included into the analysis (cf. section 2.1.2). I selected neurons within the rostral ganglia (anterior, dorsal, lateral, and ring ganglion) for optimal component placement analysis (133 neurons; 746 unidirectional connections made by chemical synapses). Furthermore, I analyzed wiring of the total global network (256 neurons; 1,751 unidirectional connections).

3.4.2 Methods and Results

In a first approach I examined permutations of the position of the 95 areas in the cortical network. As an exhaustive search of all $95! = 10^{148}$ possible position permutations was computationally unfeasible, I tested a subset of 10^7 randomly chosen permutations. For each permutation the difference between total wiring length of the permuted network and that of the original cortical network (total length: 50,455 mm) was determined. Scanning the samples, all permutations had a longer total wiring length compared to the original network with at least 7,023 mm or 14% additional length. The longest occurring wiring length was 121,766 mm, that is, 2.4 times the original wiring length. However, in each permutation at least 85 areas were placed at new positions. Therefore, it could not be excluded that smaller 'reshuffling' which only replaced 2–5 nodes might lead to lower total wiring. Such smaller changes in area positions would also be in accordance with a gradual change in wiring organization during evolution.

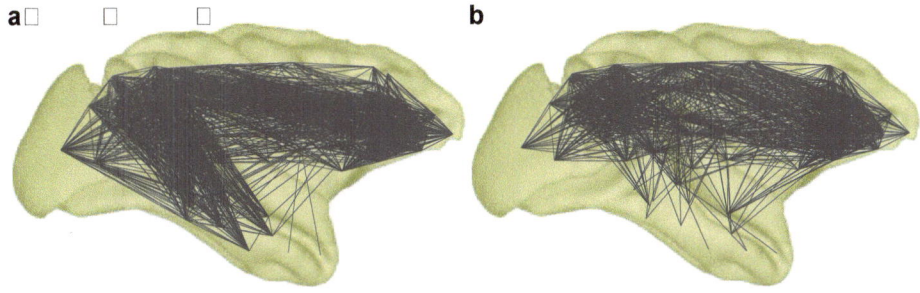

Figure 3.6: Macaque cortical connectivity network with 95 areas. **a**, Original placement of areas. **b**, Placement after evolutionary optimization (disregarding area sizes).

Therefore, I used a simple evolutionary algorithm to investigate reductions in total wiring length. Pairwise permutations of two areas (that means, swaps) were tested for lower wiring length. The first permutation that yielded a lower wiring length was kept for the next round, and once again, all two-area permutations were tested. This procedure was repeated until no permutations producing a further decrease of wiring length were identified. The algorithm exhaustively searched all possible permutations of two areas; however, it still did not consider all possible orderings of the nodes. Thus, it was not guaranteed that the global minimum was identified, so even lower total wiring lengths might be possible.

For the macaque cortical network, I found that the algorithm reduced the total wiring length by 30% of the original value (Fig. 3.6). However, one property not taken into account was the effect of different area (surface) size on the positions itself. If two areas of different size were switched, the smaller area would take the place of the larger area. In order to fill the 'empty' surface, nearby nodes would have to move closer. At the same time, the larger area that moved to the place of the smaller area would need more space so that nearby areas would be pushed further away from that position. For example, exchanging the small area LIP (lateral intraparietal area) and V1 (the largest visual area) would result in a shifting of the cortical surface influencing the positions of other cortical areas and ultimately changing total wiring length. Therefore, I limited permutations to areas whose surface size differed by not more than 5%. Although the number of permissible switches was now greatly reduced, total wiring length could still be reduced by 12.5% relative to the original value. These results demonstrate that stochastic sampling is insufficient to prove that a network shows optimal wiring. Instead, such an approach should be complemented by an evolutionary optimization algorithm searching at high resolution, that is, affecting only few areas at each step.

Using an identical optimization approach, I investigated wiring lengths within selected ganglia as well as for the whole neural network of *C. elegans*. The subset of 133 neurons within the rostral ganglia of *C. elegans* consisted of neurons of the anterior, dorsal, lateral, and ring ganglion. Here, the total wiring length could be reduced by 46% relative to the original value (Fig. 3.7a,b). Evaluating the whole neural network, which included many long-distance connections, resulted in an even larger possible reduction: by rearranging the positions of all 256 neurons

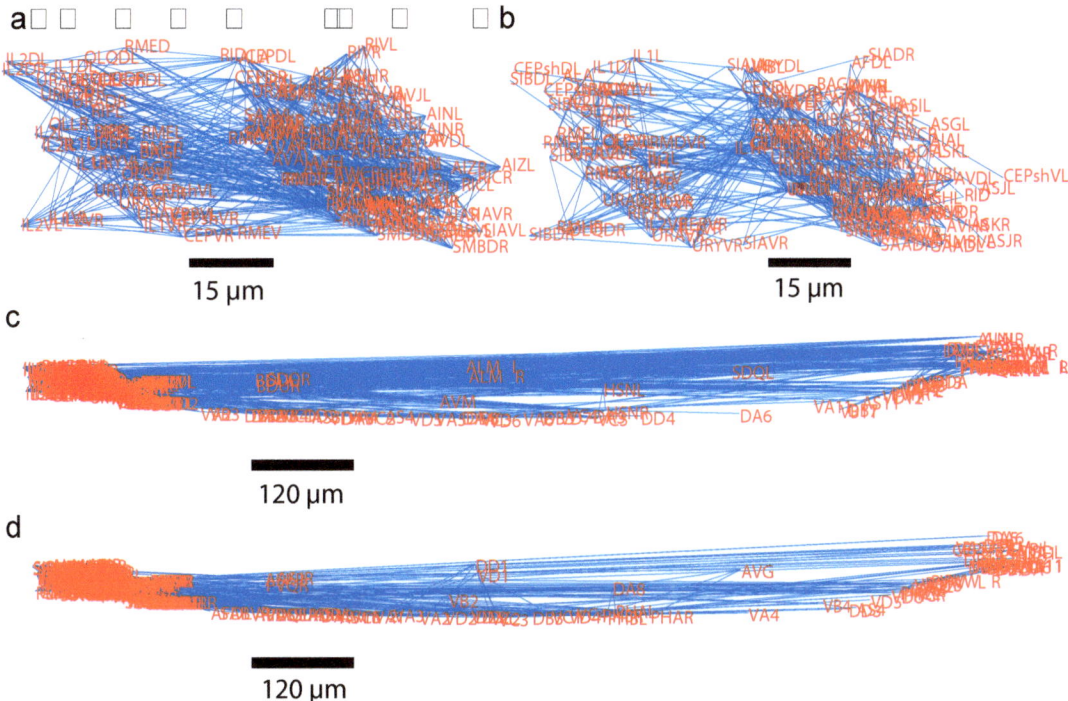

Figure 3.7: Layout of the neural network of *C. elegans*. Note that projections are represented by straight lines and could follow different paths in the animal (most long-distance connections, for example, run through the ventral cord). **a**, Neurons within rostral ganglia. **b**, Rostral ganglia neurons with alternative component placement. **c**, Global network (lateral view). **d**, Global network with alternative neuron placement.

in the network, the total wiring length could be reduced by 64% (Fig. 3.7c,d).

3.4.3 Discussion

I have presented an analysis of optimal component placement using metric distance as a measure of connection wiring length. I used a simple evolutionary algorithm to reduce total wiring length through pair-wise position switching. Both in the macaque cortical and the *C. elegans* neural network, alternative component placements with much lower total wiring lengths were found.

Using an improved methodology, the present study casts doubt on the previously reported optimal component placement in neural networks (cf. Young & Scannell, 1996 for earlier criticism). Instead of previous approaches that employed neighborhood relationships or ordinal connectivity data (Klyachko & Stevens, 2003), I used metric data for connection distances. Moreover, I analyzed networks both more extensively and at a finer level, for global cortical networks rather than subsets of areas, and individual neurons rather than whole ganglia (Cherniak, 1994).

Previous analyses (Klyachko & Stevens, 2003), for example, studied only a subset of 11 prefrontal cortical areas, calculated distances in two dimensions using a flattened map, and minimized axon volume with ordinal values for connection strengths. In addition, previous studies of *C. elegans* connectivity have analyzed ganglion positions, but not those of individual neurons.

In conclusion of these analyses, alternative constraints may shape neural networks (Kaiser & Hilgetag, 2004b; Sporns et al., 2004) that are as, or even more, important than wiring minimization.

3.5 Alternative wiring constraints

3.5.1 Introduction

The brain architecture shows a remarkable degree of organization. Surely, the architecture will influence the possible function but it remains unclear, which properties are evolutionary optimized and which are a by-product of development. In particular, the question remains which features of architecture are genetically coded, for example, by growth factors, and which are due to activity, mechanical constraints, and self-organization. One constraint of neural and cortical networks just mentioned before is a preference for minimization of wiring length.

Here I show that for the macaque cortical network there is a second property that is optimized, namely the average number of fibers that have to be crossed to go from one area to another. In addition, the cluster architecture enhances synchrony and small-world properties of the network. Therefore, the network was found optimized for low total wiring length but also for efficient processing (higher synchrony by low number of intermediate nodes).

3.5.2 Data and Methods

Data of connections and area positions.

Connectivity of 95 cortical areas in the macaque monkey with corresponding spatial positions was used for the analysis. The data set was described in section 2.1.1.

Minimal, random, regular (lattice), and rewired networks.

A minimal wiring length configuration was calculated by using original (cortical) area positions and distributing the number of found edges over the network. All possible positions for edges in the adjacency matrix were ranked by their distance so that nodes with the lowest distance between them would be connected first. This procedure was then repeated until all edges were distributed to the shortest possible distance at the respective step.

For random wiring, edges were distributed randomly (disregarding the distances between areas). Again, the positions of areas remained the same.

For regular (lattice) matrices, networks with a number of nodes and edges that was the same as for the cortical network were generated in the following way. Entries of the matrix were filled directly adjacent to the main diagonal until the number of desired connections E was reached. For example, if $E = 2N$ (N is the number of nodes), this procedure would result in nearest neighbor connectivity (ring topology) (Sporns & Zwi, 2004). These networks have high neighborhood connectivity, that means, a high clustering coefficient but lack multiple clusters.

For rewiring, pairs of edges were switched with the degree of all involved nodes remaining the same. Thus, rewiring did not change the degree distribution of the network.

Measuring synchronous activity.

In order to assess whether the cortical network was optimized for synchrony, I assumed that each cortical area introduces one neuron to the processing pathway. Naturally, this is only a lower limit of the delay that will be caused by one cortical area as multiple intermediate neurons might lie along the path through one area. However, systematic data on their number for each cortical area was unavailable. Furthermore, as global information on input patterns of cortical areas or the network architecture within cortical areas was incomplete or missing, I did not use population averages or analytical models. The applied model is therefore more of an approximation of the fastest possible ("onset") processing (Trappenberg, 2002, section 4.3.4). It will be observed to what extent neurons are able to fire simultaneously. This gives an approximation of how many areas could be activated at the same time. The actual number of neurons or areas being activated at the same time *in vivo* will be different from that predicted by this model. However, the intention of the model is to compare the simultaneous activation patterns for different network topologies rather than giving exact values of synchrony for one topology.

As each area introduces one intermediate neuron, the network comprises the minimal possible neural network over the given cortical connectivity. A neuron's connections are the same as for the represented cortical area. Whereas the model includes no interference with neurons of the same area, as only one neuron represents an area, a random input for each neuron is assumed. All synapses—as representing the terminal of long-range projections between cortical areas—are excitatory.

As model for the behavior of individual neurons—representing cortical areas—I used a simple model for spiking neurons (Izhikevich, 2003; Izhikevich, 2004). The approach is based on the passive integrate-and-fire model where the behavior of a neuron is approximated by modeling only subthreshold membrane potential dynamics without including active membrane conductances (Dayan & Abbott, 2001). The model only takes into account the network topology but not the spatial arrangement of network nodes. That means, delays that result from different axon lengths are not part of the calculation. Weights of connections were set to one for existing and zero for non-existing connections. At each time step, each node receives random input I. The change of membrane potential v of a neuron relied on two differential equations:

$$v' = 0.04v^2 + 5v + 140 - u + I$$

$$u' = a(bv - u)$$

with the auxiliary after-spike resetting

$$if\ v \geq 30\ mV, \quad then\ v \leftarrow c;\quad u \leftarrow u + d$$

with parameter values of $a = 0.02$ (time scale of the recovery variable u), $b = 0.2$ (sensitivity of the recovery variable u), $c = -65mV$ (after-spike reset value of the membrane potential), and $d = 2$ (after-spike reset of the recovery variable u). For details about the model and the role of the different parameters, see Izhikevich (2003) or check the Matlab code available at http://www.izhikevich.com.

Synchrony was measured as the average percentage of neurons that fire at the same time step, that is, among all times where one or more neurons were firing the average number of firing neurons was calculated and divided by the total number of neurons (95, corresponding to the number of areas). The synchrony of each network was assessed 100 times and the average value was assigned as synchrony of the given network. For each rewiring step, that means for each number of rewired edges, 100 networks were generated and their respective synchrony was tested.

3.5.3 Results

Cortical wiring architecture

Many long-distance connections exist in the macaque brain, for example, 9.6% of connections are longer than 40 mm (Kaiser & Hilgetag, 2004b). The longest approximated fiber length, connecting V1 and area 46, with approx. 58 mm length is not far from the longest possible distance (69 mm) for fibers between two areas within one hemisphere. In Fig. 3.8a the original macaque cortical network is shown whereas in Fig. 3.8b connections were distributed in a way that only the shortest-possible distances between areas lead to a connection between these areas. Note the absence of long-distance connections between occipital / parietal areas and frontal areas as well as between occipital and temporal regions.

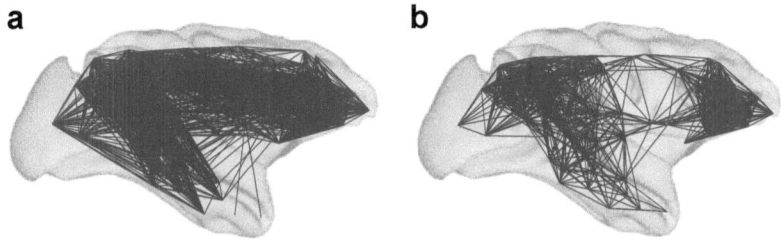

Figure 3.8: **a**, Macaque cortical connectivity with cortical areas and lines representing known corticocortical fiber connections. **b**, Minimal network generated by using the same positions but generating only shortest-possible connections.

Comparison with alternative wiring scenarios

I tested the effect of different wiring scenarios on the average shortest path (ASP) and the total wiring length (Fig. 3.9). The total wiring length of the cortical network was between random and minimal wiring. Whereas total wiring length was higher than for the minimum case, the ASP was much lower. Therefore, reducing ASP might be a constraint competing with wire length minimization.

Whereas minimal wiring yielded even higher clustering coefficients than the cortical or regular (lattice) configuration (Tab. 3.3), the ASP was much larger than for cortical or random networks. Indeed, using minimal wiring, the small-

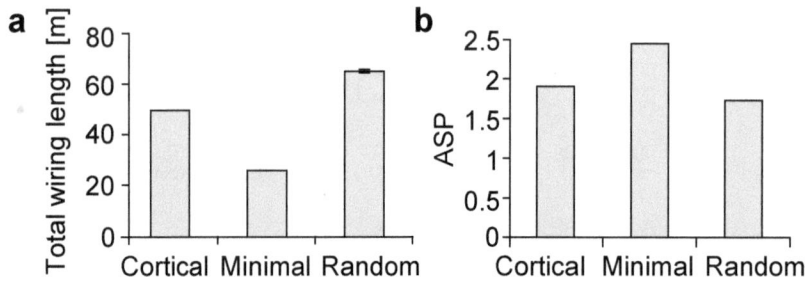

Figure 3.9: **a**, Average shortest path for cortical and minimally and randomly wired network. **b**, Total wiring length for cortical and minimally and randomly wired network.

Table 3.3: Values for ASP, C and total wiring length of cortical and alternative wiring arrangements.

	Lattice	Random	Minimal	Cortical
ASP	2.80	1.73	2.43	1.90
C	0.762	0.271	0.767	0.652
Total wiring length [mm]	66,620	66,566	26,924	50,455

world properties of cortical networks (Watts & Strogatz, 1998; Hilgetag & Kaiser, 2004) were lost.

Loss of cluster architecture

Are these properties of low ASP related to the cluster architecture of cortical networks or only related to the degree distribution? In order to answer this question, I looked at rewired networks where connections between areas were permuted while the degree of cortical areas remained the same. This blurred the defined cluster architecture as relatively more connections between clusters occurred. As the degree distribution remained the same, the effect of cluster architecture could be tested. I rewired up to 3,000 edges in steps of 100 and measured ASP, clustering coefficient

C, synchrony and total wiring length (at each step, I showed the mean value for 50 rewired networks). The values (except for synchrony) were normalized to the minimal-random scale in the following way to give relative measures (following Sporns & Zwi, 2004):

$$a_{scl} = (a - a_{random})/(a_{minimal} - a_{random})$$

$$c_{scl} = (c - c_{random})/(c_{minimal} - c_{random})$$

The ASP before rewiring was already close to the value for random networks and got closer during rewiring (Fig. 3.10). The relative clustering coefficient, however, was close to minimal wiring ($\sim$0.8) before rewiring but was reduced by 50% during rewiring (Fig. 3.10b). Therefore, the high clustering coefficient as found in small-world networks did not remain for the rewired macaque network.

Synchrony was also reduced even after only few edges were rewired (Fig. 3.10c). This could be due to the high connectivity within clusters. As many nodes within clusters were directly connected to other nodes in the cluster, activity starting at one node could quickly activate the entire cluster. Therefore, the whole cluster could be active at the same time resulting in higher overall synchrony.

In addition, the rewired network exhibited a larger total wiring length (increase of more than 30%, Fig. 3.10d). Therefore, the cluster organization helped to ensure both small-world properties and low total wiring length.

Interestingly, the decrease in ASP and synchrony to the final value already occurred for only 300 rewired edges. On the other hand, the decrease of the clustering coefficient remained constant only after 1,000 rewired edges. Therefore, synchrony seems to be linked primarily to ASP rather than neighborhood connectivity as

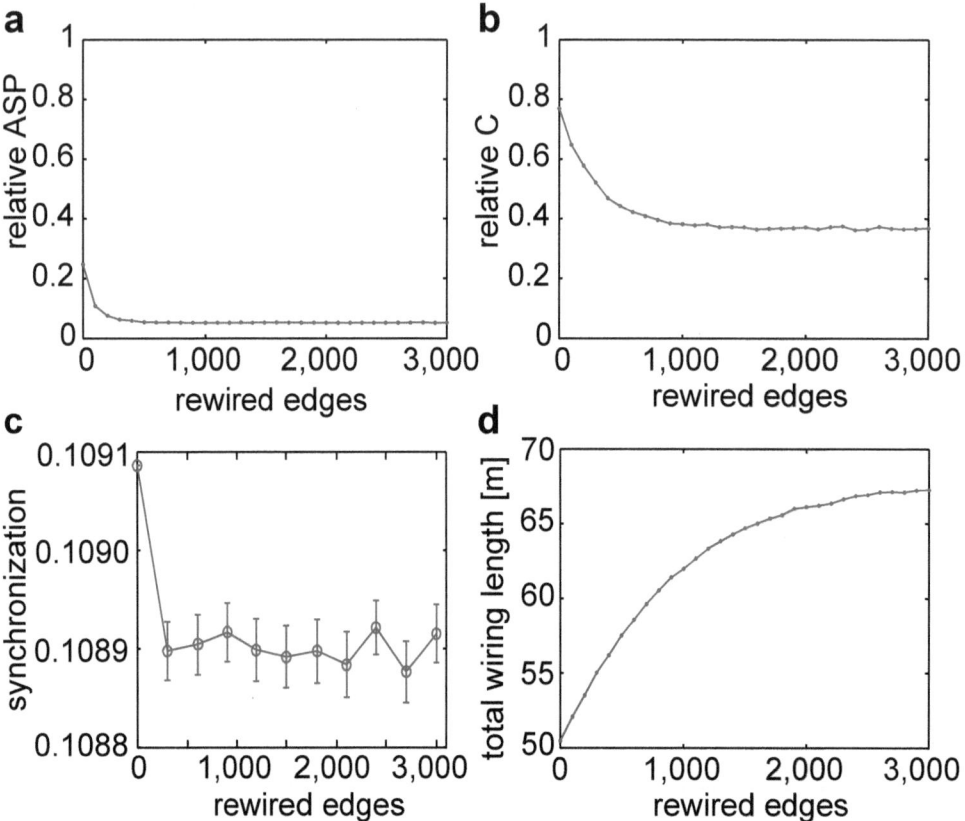

Figure 3.10: Relative (scaled) values after rewiring the macaque cortical network of 95 nodes. The degree distribution of the network was preserved. **a**, Relative average shortest path (a_{scl}). **b**, Relative clustering coefficient (c_{scl}). **c**, Synchrony (error bars indicate standard error of the mean). **d**, Total wiring length.

measured by the clustering coefficient. In addition, more rewired edges (>2,000) still increased total wiring length even when other topological values remained the same.

3.5.4 Discussion

Constraints of brain architecture

The existence of long-distance connections in the macaque cortical network helped to minimize the ASP whereas the total wiring length was larger than for the minimum scenario. Why would ASP minimization be so important to justify the effort of establishing costly long-distance connections? There are two possible reasons for this. First, each additional intermediate area means at least one neuron in the transmission line adding a delay to the processing. With only short-distance connections it would be impossible to have synchronous activity in neighboring and faraway areas that are linked over many intermediate nodes (see Singer, 1993 for a review on temporal aspects of cortical processing). Second, each additional neuron in the line would add unwanted input to the signal, which would be prevented by a direct (long-distance) connection.

The ASP for minimal wiring was closer to the ASP of a regular (lattice) network. This is reasonable as for minimal wiring, only direct neighbors on the surface rather than indirect neighbors and faraway areas were connected; therefore, connections mostly run along the surface similar to a two-dimensional lattice arrangement.

The role of the cluster architecture

A different property related to neighborhood connectivity and total wiring length was the cluster architecture of the macaque brain. Whereas it resulted in slightly higher ASP compared to the rewired networks, the total wiring length was reduced. In addition, clustering coefficient and synchrony were much larger before rewiring.

Another advantage of clusters was the increased matching of incoming and outgoing connections between areas (Passingham et al., 2002). The matching index (Sporns, 2002) of two cortical areas is defined as the average percentage of identical incoming or outgoing connections of the two areas. In the following, I looked at the average matching index over all possible pairs of areas. For the macaque cortical network, on average 13.9% of the connections of two areas had identical projection targets or sources. This was close to the 16.6% of a regular network. For the rewired network (3,000 changed edges), however, the matching index was with 6.8% only half of the cortical value (the value for a random network is 4%). The higher matching index for cortical networks could have two effects. First, as more areas have similar input and output, similar functions can be realized on a large scale. Second, the higher matching index could enhance the probability of functional compensation after brain lesions. With a high matching index, areas similar to the lesioned area would exist that could take over the function.

Apart from functional cooperation, clustering may achieve a balance between functional segregation and integration, resulting in functional connectivity of high complexity (Sporns et al., 2000b), while conserving wiring length. In addition, the close association of areas within clusters lends itself to efficient recurrent processing. Closed feedback loops among areas are very likely to occur, given the high frequency of reciprocal connections (Felleman & van Essen, 1991; Kötter & Stephan, 2003; Sporns et al., 2000b) and abundance of short cycles (Sporns et al., 2000b) in cortical systems.

Possible developmental mechanisms

Which biological mechanisms could yield small-world properties and low total wiring length? As will be discussed in the following chapter, distance-dependent connection establishment (Kaiser & Hilgetag, 2004c), where the probability to form a projection to an area decreases with the distance between source and target region, could be one approach. Distance-dependence could be due to the diffusion of growth factors that establish a concentration gradient depending on the distance to the source of the growth factor (Murray, 1990). By such a distant-dependent model it is possible to reproduce the global properties of cortical networks (Kaiser & Hilgetag, 2004a).

3.6 Summary

In this chapter, properties of cortical network organization were analyzed. Cortical networks exhibited some highly-connected nodes and were in many aspects similar to scale-free networks (cf. section 3.2). I will evaluate the functional effect of such architecture in the chapter on robustness. A detailed discussion on why and how a scale-free cortical organization could have developed is then given in the general discussion chapter (section 6.3).

Neural connectivity—especially over long distances—consumes a considerable amount of resources. Static resource consumption is composed of the costs for the establishment of fiber connections, for example, costs for axon growth, synapse establishment and formation of a myelin sheath around axons. These costs are independent from brain activity. On the other hand, there is dynamic resource consumption in terms of the energy that is needed to transmit action potentials over fiber projections and to re-establish the resting potential. As long-distance projections use large static and dynamic resources, it was argued that these connections only exist when no alternatives are possible. Specifically, positions of network nodes should be arranged in space so that every alternative position setting should lead to larger total wiring length (Cherniak, 1994).

Using spatial distance rather than indirect measures as well as using larger data sets, I have shown that the original neural network organization was non-optimal and that alternative configurations could reduce total wiring length by up to 30% (macaque) and 64% (*C. elegans*) respectively. This was due to many long-distance connections in these networks. As a reason for the existence of long-distance connections, I discussed the average number of intermediate nodes—as indicated by the intermediate number of edges yielded by the ASP—as a functional constraint. Long-distance connections were a short-cut that reduced the ASP. This could have three functional benefits: reduction of time delays, prevention of the introduction of unwanted signals ('noise') in pathways and synchronous processing in neighboring and distant nodes.

The results have shown that the current paradigm of wiring length minimization by optimal component placement does not hold and functional constraints, as indicated by the number of intermediate nodes, was also important. Further studies should examine the influence of different constraints also on the local level of connectivity *within* cortical areas. For this, it would be useful to define a measure for the contribution of the different constraints to the final architecture. After all, I expect a mixture of constraints including resource consumption, fast processing, integration of multi-modal features across functional clusters and other functional or structural aspects.

In the next chapter, I will look at possible mechanisms for network development of neural and other spatially distributed networks.

Chapter 4

Development

4.1 Introduction

During the ontogenetic development of the human brain, about 10^{10} neurons are arranged in specific layers of the cerebral cortex. Neurons differentiate both in morphology of their dendritic tree as well as their membrane characteristics so that more than 40 classes of neurons can be distinguished. Moreover, about 10^{14} synaptic connections between neurons are established with specific local targets (Kozloski et al., 2001; Binzegger et al., 2004). On the global scale, large-scale fiber bundles between cortical areas as well as sub-cortical regions are established leading to the characteristic cluster architecture described in the previous chapter. How can a bunch of progenitor nerve cells organize and form such distinct network architecture?

The study of the development of the nervous systems is not only interesting in itself but also in the light of possible applications. Understanding the mechanisms of normal as well as altered brain development could lead to treatments of brain dysfunctions that are caused by developmental disorders. Furthermore, in the area of artificial neural networks, analogous developmental methods could be applied to generate self-organizing networks with efficient parallel processing and multi-modal input (for example, combined processing of auditory and visual patterns). For example, whereas artificial neural networks such as recurrent Echo State Networks (Jaeger & Haas, 2004) incorporate features of neural topology, mechanisms for network development still differ.

4.1.1 Developmental factors for neural fate

Which mechanisms influence neuronal fate? The decision of proneural cells to differentiate into neural or non-neural cells was first studied in the fruit fly *Drosophila melanogaster*. In each cell there is an interaction between two cell-surface proteins, one functioning as ligand and the other as respective receptor. Receptor activation leads to a positive feedback loop. A high receptor activation results in a non-neural cell fate (Kandel et al., 2000, p. 1044).

During development, the survival of neurons is influenced by a neurotrophic factor that is secreted by other neurons and prevents cell death. This mechanism prevents an over-activation of neurons as a high activity of a target neuron results in low secretion of neurotrophic factor and thereby cell death of the neuron that projects to that target. The reduced number of incoming connections then results in a lower (more normal) target activity (Kandel et al., 2000, p. 1055). The survival of "useful" cells, however, is not limited to neurons. Indeed, all cells are destined to programmed cell death (apoptosis) unless the process is actively inhibited.

4.1.2 Factors determining neural projection targets

Besides neuron survival, the fate of neural projections determines the network topology. How are long-distance projections, for example, from the retina to the lateral geniculate nucleus and further to the primary visual cortex, established? Historically two streams of thought can be distinguished. First, neuroanatomists believed that mechanical guidance (called stereotropism) and functional activity would determine long-range connectivity. Axon growth was thought to be more or less random and the survival of connections would be determined by the matching of activity patterns ('resonance') between axon and target (Weiss, 1941). Subsequently, neural growth factors were discovered and led to the chemospecificity hypothesis (Sperry, 1963). Here, the direction of axonal growth cones to their synaptic target can be changed by various molecular guidance cues (Sur & Leamey, 2001). One well-understood example is the subcortical projection from the retina to the optic tectum. Whereas gradients can serve as long-range cues, ephrins are positional local cues at the target region. Guidance cues include extracellular matrix adhesion or cell surface adhesion, fasciculation (following existing pioneer fibers), chemoattraction or -repulsion and contact inhibition (Kandel et al., 2000, p. 1071f).

Research nowadays is mostly focused on such molecular developmental factors. Indeed, early work on mechanical constraints of fiber growth (His, 1888) or skull formation (Thompson, 2004) is unknown to many modern neuroanatomists. However, molecular factors might not explain all characteristics of development. First, transient connections occur that are removed at later stages of development suggesting that elements of trial-and-error or self-organization do exist. Second, it can be projected from known macaque connectivity that at least 1,000 corticocortical fiber bundles in the human brain exist. It seems unlikely that numerous guidance cues for each specific projection are genetically encoded.

In the previous chapter it was mentioned that most corticocortical connections are short-distance connection (Kaiser & Hilgetag, 2004b). A similar pattern can be seen on the local level of neural connections within areas (Hellwig, 2000). Therefore, I analyze how much of the cortical connectivity can be explained by a developmental algorithm in which connection establishment depends on the distance between cortical areas. Furthermore, I will evaluate how time windows for synaptogenesis (Rakic, 2002) can ensure the generation of multiple clusters.

4.1.3 Development of spatial networks

Spatial networks are graphs that consist of nodes arranged in metric (usually two-dimensional) space (Watts, 1999). However, the concept of space could be extended to non-metric space. A typical example is transportation networks where cities are the nodes with a two-dimensional position on the map and highways are connections between the nodes. In theoretical studies, spatial graphs are usually generated by distributing the nodes randomly in space. After the nodes are arranged, edges between the nodes are added to the network either randomly or depending on the distance between the nodes.

These models, however, fail to reproduce many features of real-world networks.

One such feature is that of a small-world network (Watts & Strogatz, 1998) in which many connections between neighbors of a node exist but characteristic path length, that means the number of edges that have to be crossed to go from one node to another, is as short as that of a randomly connected network. Neighborhood connectivity is usually assessed by the *clustering coefficient* which is the average percentage of connections between neighbors of a node (Watts & Strogatz, 1998). Small-world properties were found in many biological and artificial networks, including networks extending in space such as the Internet, power grids, neural networks, and food webs (Watts & Strogatz, 1998). Another property is that of having multiple clusters, that means, that nodes as well as connections between nodes are not distributed equally among space but concentrated on certain regions. Thus, there exist multiple clusters in the network. For example, in the yeast protein-protein interaction network there are relatively more interactions within the mitochondrium than between mitochondrium and other cellular compartments (Schwikowski et al., 2000). Another property of real-world networks is that highly-connected nodes or hubs exist in the network which would not occur when a random distribution of connections is assumed. The degree distribution of these real-world networks, where degree is the number of edges that a node has, follows a power-law where the probability for a node having degree k is $P(k) \propto k^{-\gamma}$ (Barabási & Albert, 1999).

I will describe a spatial growth model that can explain features of real-world networks such as scale-free degree distribution, small-world properties, optimal component placement and multiple clusters. The chapter consists of the following sections. First, I will give examples for spatial graphs and their properties. Next, I will give an overview of different models for network development or evolution. Then, I will give general properties of the new spatial growth model and will—as a case study for real-world networks—describe how brain-like networks can be generated. Finally, I will summarize the results on spatial growth.

4.2 Analyzed networks as examples for spatial graphs

Internet

One spatial network with well-documented topology is the network of networks—the Internet. Recently, a combination of spatial constraints (Waxman, 1988) and preferential attachment (Barabási & Albert, 1999) has been shown to yield a topology similar to the Internet (AS level) (Yook et al., 2002). However, all nodes were existing before the start of establishing connections thereby not modeling growth in terms of an increasing number of nodes. Here, I analyzed a snapshot of the Internet at the autonomous systems level (see section 2.2.2 for a description of the data).

Traffic networks

Cities and roads between them are geographically distributed. It was shown recently that the German highway network has a scale-free topology (Kaiser & Hilgetag, 2004c) [1] (cf. section 2.2.1 for a description of the network). As for the Internet, the density of the highway network of 0.18% is very low (Fig. 4.1a). Because of linear chains of nodes, that means, sequences of nodes that have only connections to successors or predecessors in the chain, the clustering coefficient (0.12%) is lower than the density. In addition, the maximum degree of a node (12 highways) is very low so that no highly-connected nodes exist in the system. A similar type of organization was also found for scale-free protein-protein interaction networks (Jeong et al., 2001) (maximum degree $k_{max} \approx 20$). Therefore I term these networks with linear chains of nodes and power-law but hub-less degree distribution *linear scale-free networks*.

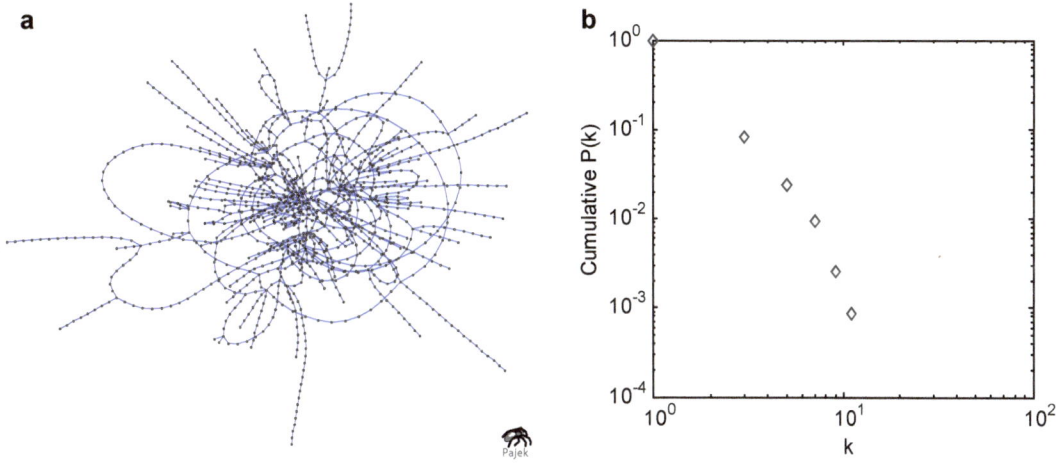

Figure 4.1: German highway (Autobahn) system. **a**, Positions were yielded by the free Kamada-Kawai energy minimization (Pajek software package). Linear chains of nodes can be clearly seen in the periphery of the visualization. **b**, Cumulative degree distribution ($k_{max} = 12$).

---

[1]In this study nodes are defined as highway exits. In other studies (Gastner & Newman, cond-mat/0407680) only highway junctions and state borders were defined as nodes. This could explain why these networks did not show a scale-free degree distribution.

Brain Connectivity Networks

As other biological systems, the brain is a spatial network. I analyzed the network not on the level of single neurons but on the level of large-scale structural modules (cytoarchitectonically defined areas) in the cat and the macaque (cf. section 2.1.1). In contrast to the previous networks, the brain connectivity networks were very dense (15% density for the macaque with 73 areas and 30% for the cat with 55 areas). The network showed properties of small-world networks, in that the clustering coefficient (Watts & Strogatz, 1998) was higher and the average shortest path (ASP) comparable to similar random networks.

In addition to connectivity networks used in previous articles (Young, 1993; Hilgetag et al., 2000), I analyzed a macaque connectivity network together with the corresponding positions of the cortical areas (95 areas, cf. section 2.1.1).

Biochemical Networks

I analyzed metabolic and protein-protein interaction networks. Whereas the edge density was only 1% in these systems, the clustering coefficient was 60% and 70%, respectively. I used data from metabolic networks of 43 organisms where the nodes were metabolites and the edges were transitions or reactions (data from `http://www.nd.edu/~networks/database/`). The protein-protein interaction network of yeast (*S. cerevisiae*), on the other hand, consisted of proteins and interactions between them (cf. section 2.1.3).

4.3 Methods for modeling network growth

4.3.1 Distance independent growth models

Growth and preferential attachment

The standard model for generating scale-free networks uses growth and preferential attachment (Barabási & Albert, 1999). Starting with m_0 initial nodes, a new node establishes a connection with an existing node i with the probability

$$P_i = \frac{k_i}{\sum k}$$

that is, the number of edges of node i, that is, k_i divided by the total number of edges yet established in the network. The resulting network consists of one cluster, and the degree distribution shows a power-law. Whereas this model can reproduce the degree distribution of many real-world networks, some topological features are not matched. For example, the model leads to eigenvalues and eigenvectors with different patterns compared to those of scale-free protein-protein interaction networks (de Aguiar & Bar-Yam, 2005). However, hierarchical modular networks can reproduce these features.

Hierarchical model

In order to develop modular scale-free networks, the hierarchical model for network generation has been developed (Barabási et al., 2001; Ravasz et al., 2002). Starting with one root node, for each step two units that are identical to the network generated in the previous iteration, are added and the bottom nodes of these two units are linked with the root of the network.

In contrast to real-world networks, the modules are generated in a deterministic way. This also means that the module hierarchy is strongly self-similar, leaving no space for a variety of module sizes within the same network. However, for metabolic networks it was argued that the replacement of a unit by the whole network, thereby increasing the previously existing topology, could be the result of copying and reusing existing modules and motifs (Ravasz et al., 2002; Reigl et al., 2004).

4.3.2 Distance dependent growth

Preference for short-distance connections

Waxman (1988) proposed a connection establishment algorithm for the Internet in which the probability of a connection between two nodes decays exponentially with the spatial distance between them. In that way, the high costs for the wiring and maintenance of long-range connections can be represented.

The nodes are distributed at random. Thereafter, edges are attached to the graph. The probability that an edge is established between two nodes u and v is

$$P(u,v) = \beta \, e^{\frac{-d(u,v)}{L\,\alpha}}$$

where L is the maximum distance between two nodes, α is a factor for the distance importance (a higher value results in fewer long-range connections), and β is a scaling factor determining the desired edge density of the network. The probability decays exponentially with the distance $d(u,v)$ between the nodes. The location of the nodes is determined from the start, therefore, there was no growth in terms of the size of the network as given by the number of nodes.

Diffusion as a reason for distance dependence in biological systems

Why should short-distance connections be more likely than long-distance connections in biological systems? One explanation is that the concentration from the place of the production / emission of a growth factor to a distant location is decreasing. The process of diffusion, building up a gradient towards the source of a molecule, is dependent on its molecular mass, the viscosity of the fluid, and the temperature. Looking at a system where a certain concentration of molecules is deposited at position $x = 0$ for the starting time $t = 0s$, the concentration at position x after time t is given by

$$c(x,t) = \frac{Q}{2\sqrt{\pi D t}} e^{\frac{-x^2}{4 D t}}$$

where D is the diffusion coefficient and Q is the initial amount of particles per area (Murray, 1990, p. 235).

4.3.3 Spatial growth model

The diffusion model is particularly plausible for biological networks, as the concentration of growth and signaling factors decays exponentially from the point of origin in the cell or the brain. Therefore, the probability for establishing a connection with a certain brain area or to interact with a certain protein will decrease with spatial distance to the growth factor source or the protein, respectively. In this new *spatial growth* model nodes are generated step by step, and whenever a new node is added to the network, its connections to the existing network are established.

This method can take borders or limits of development into account. The generation of new cortical regions is limited within the skull. In addition, apoptosis factors limit the generation of neurons so that growth beyond certain limits does not occur. In that way new nodes can not be formed at the borders of the network, that is, distant from the existing network. As new nodes can only be generated within the limits of the existing network, the core of the network is increasing its connection density. On the other hand, a network that did not reach the borders when the desired network size was generated can be called 'virtually unlimited'.

As the initial spatial distribution of nodes would have been an additional parameter for network growth, I started network generation with one defined initial node at the center of the network space. In that way the initial distances to the border for the growing network were identical for all trials.

The following procedure was then applied: At each step a new node was added at a randomly chosen position. Thereafter, for all existing nodes the probability for establishing an edge between the new node u and an existing node v was

$$P(u,v) = \beta \ e^{-\alpha \ d(u,v)} \tag{4.1}$$

where β is a scaling factor and α controls the probability for establishing long-range connections. New nodes that did not establish connections with existing nodes were removed from the network. The procedure was repeated until the desired number of nodes was reached.

In a slightly modified approach the growth model could employ a power-law to describe the dependence of edge formation on the spatial distance of nodes:

$$P(u,v) = \sigma\, d(u,v)^{-\tau}. \qquad (4.2)$$

By this mechanism the probability of establishing distant nodes would be increased even further. For example, simulating networks of similar size (50 networks; $n = 100$ nodes; $density = 4\%$; square embedding space with edge length 100) for both types of distance dependencies, the power-law (Eq. 4.2, $\sigma = 1$, $\tau = 1$) resulted in higher total wiring length (6,303) compared to networks generated by exponential edge probability (Eq. 4.1, $\alpha = 0.35$, $\beta = 1$, total wiring length 1,077 units). In the following investigations, however, I concentrated on the exponential approach outlined above, since my simulations indicated that power-law edge probability (tested parameter ranges $\sigma \in [0.004; 2]$ and $\tau \in [0.125; 64]$) was unable to yield small-world networks.

4.4 Results

4.4.1 General properties of spatial growth networks

Parameter domains for different network types

For very small β (< 0.01), sparse networks were generated (Fig. 4.2a) in which only a small proportion of all possible edges was established. The resulting networks were highly linear, that is, exhibiting one-dimensional chains of nodes, independent of limited or virtually unlimited growth (parameter α). The histograms of chain-lengths found in these networks, indicating the number of nodes in the chains, were similar to those of random networks with the same density. Unlike in random networks, however, the clustering coefficient was lower than the network density, and despite lacking clusters and hubs with large degree k, these networks possessed a power-law degree distribution, with high ASP (Fig. 4.2b, to avoid systematic errors known to occur for linearly histogrammed data plotted on logarithmic scales, the plot uses data bins of uniform width). The power-law exponent was small, in the range of $[1.7; 2.1]$; and in the simulated networks of 100 nodes the cut-off for the maximum degree of the scale-free networks was 16. Given their low maximum degree, these networks with low clustering and long linear chains of nodes were called linear scale-free.

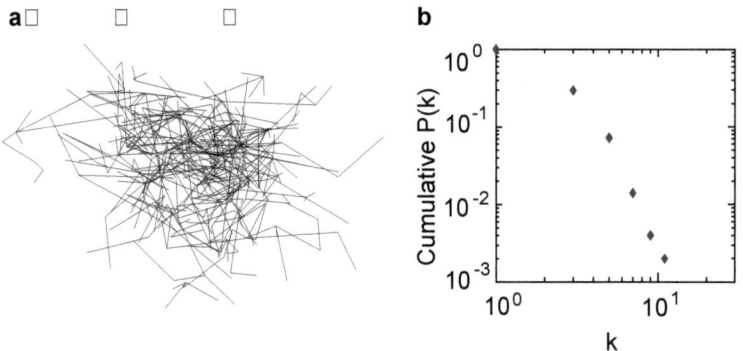

Figure 4.2: Example of a linear scale-free network generated by spatial growth. **a**, Sparse network (density 0.42%) with 500 nodes obtained by limited growth ($\alpha = 2$, $\beta = 0.001$). **b**, Double-logarithmic plot of the cumulative degree probability $P(k)$ that a node possesses k edges for the network shown in **a**. The plot is based on uniform bins of data. A power-law of the degree distribution ($\gamma = 2.43$) can be observed.

For higher edge probability ($\beta \to 1$), a noteworthy difference between limited and virtually unlimited growth became apparent. While it was impossible to generate high network density under virtually unlimited growth conditions, the introduction of spatial limits resulted in high density and clustering, as well as low ASP. This was due to the fact that, in the virtually unlimited case, new nodes at the borders of the existing network were surrounded by fewer nodes and therefore formed fewer edges than central nodes within the network. In the limited case, however, the network occupied the whole area of accessible positions. Therefore, new nodes could only be established within a region already dense with nodes and would form many connections.

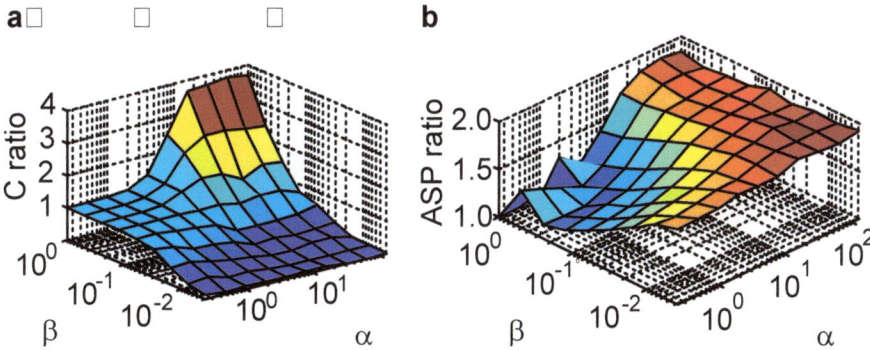

Figure 4.3: Comparison of small-world properties of spatial and random networks for $N = 100$ nodes. Each data point represents the average for 50 networks. **a**, Ratio of the clustering coefficient C of the generated networks divided by the clustering coefficient for comparable random networks. A large ratio is one feature of small-world networks. **b**, Ratio of the average-shortest paths, ASP, of spatial-growth and comparable random networks.

Figure 4.3 shows the relation between small-world graph properties and growth parameters α and β for networks consisting of 100 nodes. The ratio of the clustering coefficient in spatial growth compared to random networks was larger than one (indicating small world graphs), when the values for α and β were high (Fig. 4.3a). The ASP in the generated networks normalized by the ASP in random networks with similar density was similar for low values of α and high values of

β. For these networks the likelihood of edge formation was high and—because of the low value of α—independent from spatial distance. Such networks resembled random growth, with the clustering coefficient possessing the same value as the density ($C/C_{random} \approx 1$).

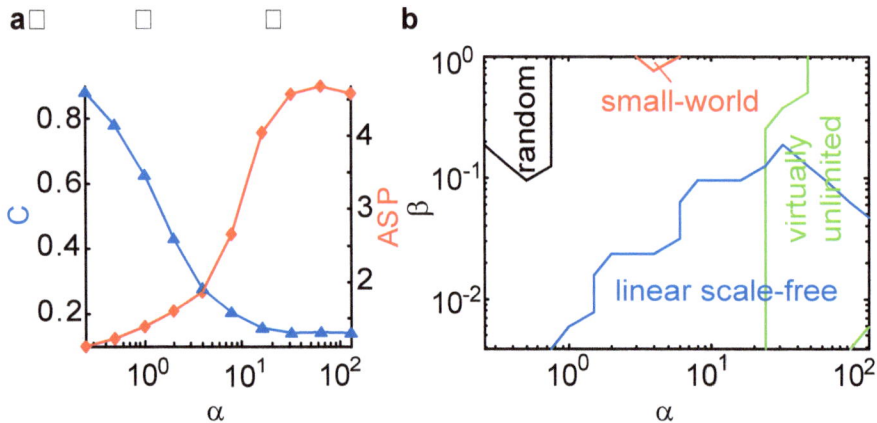

Figure 4.4: Exploration of model parameter space. **a**, For dense networks ($\beta = 1$, $N = 100$ nodes), an increased dependence of edge formation on distance (parameter α) led to an increase of ASP (diamonds) and a decrease in clustering coefficient C (triangles). **b**, Overview of network types for different spatial growth parameters ($N = 100$ nodes). Low values of α made edge formation independent from distance and resulted in random networks. For large values of α only nodes near the existing network could establish connections, and the hard borders were not reached (virtually unlimited). The area labeled linear scale-free was a region in which networks were sparse and highly linear and showed a scale-free degree distribution occurred. Only a small part of the parameter space displayed properties of small-world networks.

In a small interval of intermediate values for α ($\alpha \approx 4$, $\beta = 1$), networks exhibited properties of small-world networks (ASP and clustering coefficient shown in Fig. 4.4a). Here, the ASP was comparable to that in random networks of the same size ($\lambda \approx \lambda_{random}$), while the clustering coefficient was 39% higher than in random networks (Watts, 1999, p. 114). An overview of the parameter space and the resulting random, small-world, virtually unlimited or linear scale-free networks is given in Figure 4.4b.

In contrast to limited growth, virtually unlimited growth simulations with high β resulted in inhomogeneous networks with dense cores and sparser periphery. It is difficult to imagine realistic examples for strictly unlimited development, as all spatial networks eventually face internal or external constraints that confine growth, may it be geographical borders or limits of their energetic and material resources. However, virtually unlimited growth may be a good approximation for the early development of networks before reaching borders.

I also tried an alternative approach for generating scale-free networks. The idea was to increase the probability that new nodes settle in dense areas inside the network rather than at positions where only few nodes existed. Therefore, a node only survived if it established more connections than the average connectivity. A node inside a cluster had more links than a node at more sparse locations of the network. Consequently, its probability to survive was higher. The resulting network covered a smaller area of the network space, as it was more likely that nodes survived at the core of the network rather than at its borders (Fig. 4.5). The density was increased and the ASP for the unlimited case is now similar to the limited case. However, no hubs occurred and no scale-free degree distribution could be generated. As new nodes inside clusters had the same probability to establish links with peripheral as well as central nodes, the increase in connectivity did not favor a small number of nodes and did not result in hub formation.

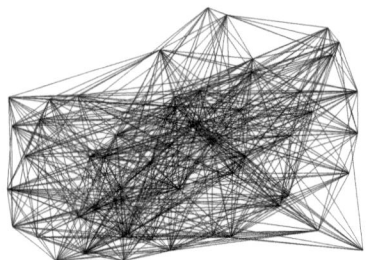

Figure 4.5: Preferential crowding condition for limited growth. The core of the network has a much higher node density than the periphery. However, no power-law degree distribution or individual highly-connected nodes were found.

4.4.2 Distinguishing types of network development

Different network growth types could be distinguished by assessing the evolution of network density and clustering coefficient. Growth with preferential attachment as well as spatial growth led to clustering coefficients, $C(N)$, that depended on the current size of the network, that is, the number of nodes, N (Fig. 4.6a). While $C(N)$ decreased with network size for networks generated by the BA-Model (Barabási & Albert, 1999), it remained constant for spatial-growth networks. Virtually unlimited or limited spatial growth could thus be distinguished, since density decreased with network size for unlimited growth, while remaining constant for limited growth (Fig. 4.6b).

Example I: Evolution of metabolic networks. I applied this concept to classifying the development of real-world biological networks. The evolution of metabolic systems, for instance, can be seen as an incorporation of new substances and their metabolic interactions into an existing reaction network. Reviewing 43 metabolic networks in species of different organizational level (Jeong et al., 2000), the clustering coefficient of these systems remained constant across the scale

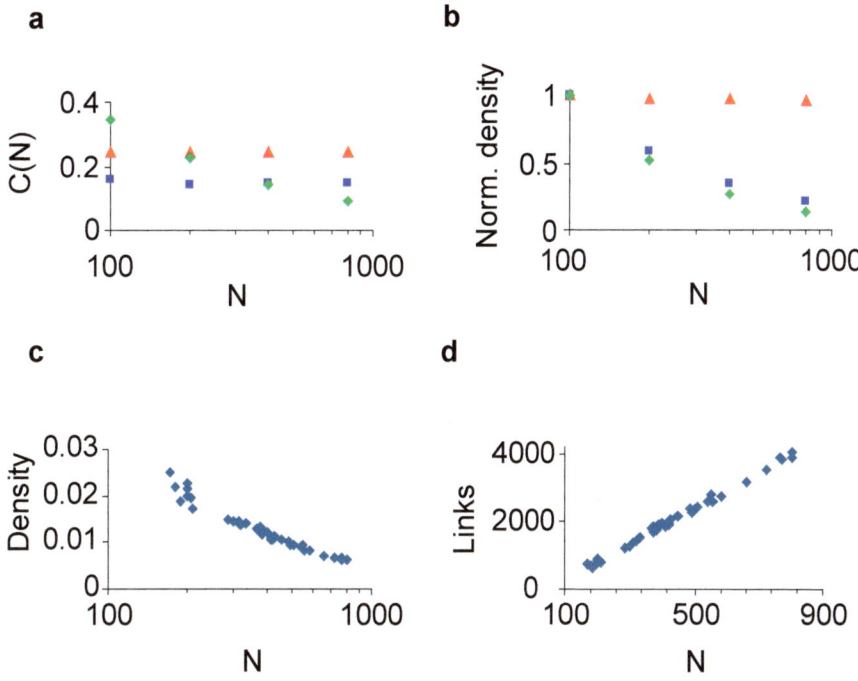

Figure 4.6: Comparison of the dependence of clustering coefficient $C(N)$ and density on network size (number of nodes, N). **a**, For the simulated networks the clustering coefficient remained constant for limited (triangles, $\alpha = 5$, $\beta = 1$) and virtually unlimited (boxes, $\alpha = 200$, $\beta = 1$) spatial growth, but decreased for growth with preferential attachment (diamonds). **b**, Density was independent of network size only for limited spatial growth. **c**, Density depending on network size (N) for the metabolic networks of 43 different organisms (15). **d**, A critical measure for network development was the dependence of network size on the number of links. For metabolic networks, this relationship was strongly linear.

(Ravasz et al., 2002), whereas their density (Fig. 4.6c) decreased with network size. This indicated features of virtually unlimited network growth. The relation between the number of links and nodes in these systems was linear (Fig. 4.6d), with a slope of 5.2, so that the number of interactions of a metabolite was not increasing with network size. Such linear growth may ensure that the metabolic systems remain connected (with the number of reactions larger than substances, as a necessary condition for connectedness), while not becoming too complex too quickly (as, for instance, with exponential addition of new reactions).

Example II: Evolution of the Internet. The development of the Internet was considered at the autonomous system (AS) level, that is, for independent subnetworks. Over the years, the number of autonomous systems has been increasing. Whereas the clustering coefficient did not shrink and even seemed to increase, the density has been decreasing with network size (Fig. 4.7). Therefore, the current development of the Internet shows properties of unlimited spatial growth. Indeed, current maps still show sparsely populated countries to which the network can expand (Yook et al., 2002, Fig. 1). In order to determine the exact growth mechanisms, it would be interesting to investigate to what extent new nodes form at dense geographical regions, that is, within the network, or at sparse regions at the network borders.

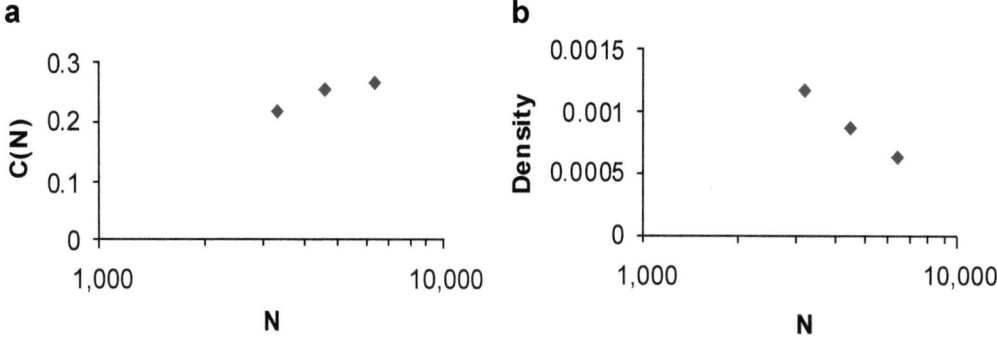

Figure 4.7: Evolution of the growing Internet. Clustering coefficient $C(N)$ and density depending on the size of the AS-network. **a**, After an increase from 1997 to 1998 (first two data points) the clustering coefficient remained constant. **b**, Density of the network was decreasing with network size.

I subsequently applied the spatial growth model to cortical network development.

4.4.3 Case study: cortical connectivity networks

Here, I analyze cortical networks of the cat and the macaque monkey. I will test whether properties of these networks can result from spatial growth. First, I will test whether spatial growth limits are necessary to generate networks that are similar to cortical networks.

The role of borders for development

The connectivity networks of cat and macaque are compared with generated spatial growth networks under the limited (border) as well as the virtually unlimited condition. The parameter β was not changed, therefore, the density and clustering coefficient differed between bordered and unlimited case. Both for the cat (Tab. 4.1) and the macaque (Tab. 4.2), limited growth could yield similar network properties of clustering coefficient and ASP, but the virtually unlimited condition led to larger ASP (almost a factor of 2) and decreased clustering coefficient (about 2/3 of the original value).

It appears to be impossible to generate a high density as in the cat network and low ASP's as for cat and macaque network without borders for spatial growth. In fact, the ASP for the unlimited networks is much higher than for comparable random networks (ASP 2.8) excluding them from being small-world networks. It appears that small-world networks, indeed, only occur in *small* (limited) and not in infinite network spaces.

Table 4.1: Cat brain connectivity network and artificial networks generated by spatial growth in order to yield similar properties (both in the limited border condition as well in the virtually unlimited case). Clustering coefficient cl, density d, average shortest path ASP and the maximum degree of the resulting network are shown.

	cl	d	ASP	max. deg.
Cat$_{55}$	0.55	0.3	1.82	31
border ($\alpha=5$)	0.5	0.34	1.7	29
unlimited ($\alpha=200$)	0.3	0.08	3.9	10

Table 4.2: Macaque brain connectivity network and artificial networks generated by spatial growth in order to yield similar properties (both in the limited border condition as well in the virtually unlimited case). Clustering coefficient cl, density d, average shortest path ASP and the maximum degree of the resulting network are shown.

	cl	d	ASP	max. deg.
Macaque$_{73}$	0.46	0.16	2.2	39
border ($\alpha=8$)	0.4	0.17	2.2	23
unlimited ($\alpha=200$)	0.3	0.07	4.1	12

Total wiring length in cortical and simulated networks

I compared the wiring length of networks generated with the spatial growth model in three dimensions with the wiring length of cortical systems connectivity in one hemisphere of the macaque monkey (for information about connectivity and spatial positions of areas cf. 2.1.1).

For the macaque cortical network with 95 nodes as well as comparable generated networks (length scaled to account for different embedding space), the total length of all connections was comparable (macaque: 32,364 mm; spatial growth simulation: 36,562 mm).

Interestingly, networks generated by spatial growth show optimal component placement (Cherniak, 1995), in that reordering node positions leads to an increase in total wiring length. In contrast to previous approaches to yield optimal wiring, such as the vector mechanics model (Cherniak et al., 1999), spatial growth does not need *a posteriori* optimization. In favoring nearby nodes for connection establishment during each individual step, the greedy spatial growth algorithm leads to the global property of optimal component placement.

Multiple clusters

By using the aforementioned algorithm multiple clusters can (but may not) occur. When a new node faraway from the existing network does survive, candidate nodes in its vicinity are likely to establish a connection to such a *pioneer node*. However, the probability that a node distant to the existing network can survive in the first place, that is, establishes at least one long-distance connection to the existing network, is still low. Therefore, it can not be guaranteed that such a node occurs during development and that multiple clusters arise. Furthermore, there is no control over the size of occurring clusters.

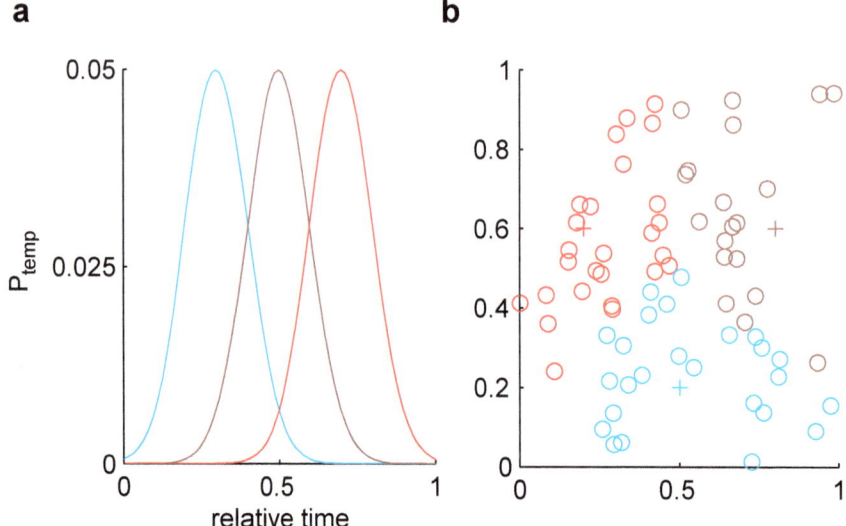

Figure 4.8: Time windows and initial seed nodes. **a**, Temporal dependence P_{temp} of the projection establishment depending on the time and the type of the node. Relative time was normalized in that 0 stands for the beginning of the development and 1 for the end of network growth. The three seed nodes had different time windows which were partially overlapping. **b**, Two-dimensional projection of the 73 three-dimensional node positions. The color represents the time window corresponding to one of the three seed nodes (+).

Therefore, I introduced a different factor that can ensure the occurrence of multiple clusters: *time windows* for development. In neural systems, time windows have been found during cortical development (Rakic, 2002). The formation of many cortical areas overlaps in time but ends at different time points with highly differentiated sensory areas (for example, area 17) finishing last. Based on this finding, I explored a wiring rule by which network nodes were more likely to be connected if they developed during the same time window.

I used the following algorithm for growth depending on distance as well as time windows (cf. Fig. 4.8a). First, three seed nodes were placed at spatially distant locations (cf. Fig. 4.8b). New nodes were placed randomly in space. The time window of the new node was the same as that of the nearest seed node as it was assumed to originate from that node. Second, the new node u established a connection with an existing node v with probability $P(u,v) = P_{temp}(u) \times P_{temp}(v) \times P_{dist}(u,v)$. Third, if no connections were established, the node was removed from the network.

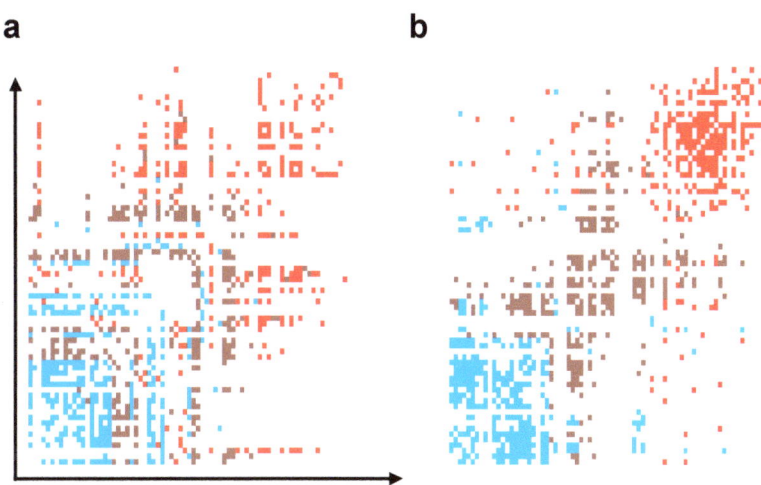

Figure 4.9: **a**, Timed adjacency matrix (the first nodes are in the left lower corner). **b**, Clustered adjacency matrix. The matrix is the some as in **a** but nodes with similar connections are arranged more adjacent in the node ordering.

The timed adjacency matrix shows the development of connections over time (Fig. 4.9a). Different colors represent the time-window of the nodes. The reordered matrix represents the original network with different node order in such a way that nodes with similar connectivity were placed nearby in the adjacency matrix (Fig. 4.9b). The resulting network exhibited both highly connected nodes and long-distance connections (Fig. 4.10). Using time windows generated a cer-

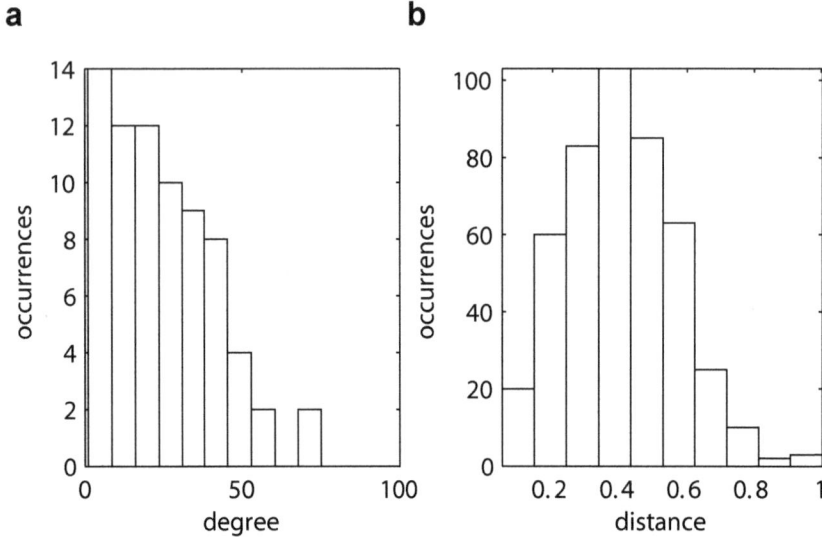

Figure 4.10: Network properties after spatial growth with time windows (73 nodes, 832 arcs). **a**, Degree distribution. Whereas most nodes had fewer than 30 (incoming or outgoing) connections, also highly connected nodes did exist. **b**, Distribution of 'fiber' lengths, as approximated by the Euclidean distance between the three-dimensional node positions, in the generated networks. Some long-distance connections also occurred.

tain number of clusters depending on the number of different time windows that govern development. In addition to the number of clusters, the size of clusters could be varied by changing the width of the corresponding time window.

In conclusion, spatial growth and time windows can represent various global network properties as well as a multi-cluster architecture. Although having similar general properties, the time window approach remains preliminary. More information about cortical time windows and the positions of clusters in the cortex has to be included in order to compare the simulation results with the existing cortical network.

4.5 Summary

The brain, like many other networks, extends in space. However, most current algorithms for network generation do not take into account the spatial position of nodes but rather node or edge properties of the existing network. I have shown that a simple model of spatial growth, where the probability for edge establishment between two nodes decays exponentially with their distance, can generate a large range of real world networks. In addition, networks similar to cat and macaque cortical networks could be generated under the condition that the borders of the spatial embedding space were reached quickly and subsequent node (and connection) establishment was limited to spaces within the existing network. The model of spatial growth could be also relevant to networks without metric Euclidean distances. For example, proteins in protein-protein interaction networks can be confined to reaction compartments (for example, endoplasmatic reticulum, mitochondria, or cell membrane) with higher interaction ($\hat{=}$ connection) probability for proteins within the same compartment. By using time windows in addition to distance-dependence, the number and size of clusters in the network can be adjusted. Therefore, much of the cortical topology can be generated by few simple mechanisms.

Moreover, I have presented a network growth algorithm which is able to yield networks with high correlation coefficient and low ASP as found in small-world networks by using borders for network growth. Moreover, the clustering coefficient remains independent from the network size as found for example, for metabolic networks (Ravasz et al., 2002). I was also able to show that (linear) scale-free spatial networks can be generated *without* preferential attachment, only relying on the distance between two nodes. Furthermore, networks similar to cortical networks of cat and macaque could be generated without a need for neural spiking activity.

Chapter 5

Robustness

5.1 Introduction

Extensive evidence shows that biological networks are remarkably robust against damage of their nodes as well as of their edges (Finger & Stein, 1982). For example, Parkinson disease only becomes apparent after a large proportion of pigmented cells in the substantia nigra are eliminated (Damier et al., 1999), and in spinal cord injuries in rats, as little as 5% of remaining fibers allowed functional recovery (You et al., 2003). Many metabolic networks, as well, were found to be robust against the knockout of single genes. This feature is both due to the existence of duplicate genes as well as of alternative pathways which ensure that a certain metabolite can still be produced using the undamaged parts of the network (Wagner, 2000).

Also at the global level, the brain can be remarkably robust to physical damage. Significant loss of neural tissue can be compensated for in a relatively short time by large-scale adaptation of remaining brain parts (Spear et al., 1988; Stromswold, 2000). Contrastingly, some targeted damage even to relatively restricted brain areas may lead to dramatic changes: targeted lesions, for example in the amygdala (Coover et al., 1992; Muller et al., 1997) and the hippocampus (Zola-Morgan et al., 1986), cause significant and long-lasting behavioral effects.

These findings provide a somewhat contradictory picture of the robustness of the brain and suggest a number of questions. Can we evaluate effective robustness given this variability in the effects of brain lesions? Which nodes or edges of the network are most vulnerable? Can the effect of multiple lesions be predicted from single lesions? I assess here how connectivity data of brain area connectivity can be brought to bear on these questions.

In section 5.2, I show *which* edges are most important in cortical and biochemical networks. For this, I compare different measures for predicting the damage after the elimination of individual edges. Furthermore, I will discuss *where* in the network the most important edges reside. In section 5.3, I will compare the robustness after lesions of cortical networks of cat and macaque with scale-free networks. More specifically, I will compare the damage of cortical and benchmark networks after sequential removal of nodes or edges. In section 5.4, I test whether multiple lesions can give additional information about the contribution of nodes compared to single lesions. Indeed, I find that for some combinations of nodes their elimination has a different effect as would be expected from single lesions.

5.2 Predicting the effect of single edge removal

5.2.1 Introduction

Whereas previous studies explored the impact of removing network nodes (Barabási & Albert, 1999), the effect of *edge* elimination in biological networks has not yet been investigated. How can edges that are integral for the stability and function of a network be identified? In some systems, for instance, transportation or information networks, indicators for the importance of edges, such as flow or capacity, are available. For many biological networks, however, such measures do not exist. In brain connection networks, for example, a projection between two regions may have been reported, but its structural and functional strength is frequently unknown or not reliably specified (Felleman & van Essen, 1991), and its functional capacity may vary depending on the task (Büchel & Friston, 1997). Similarly, in biochemical networks, reaction kinetics are often highly variable, or generally unknown (Schuster & Hilgetag, 1994; Stelling et al., 2002). However, the analysis of a network's structural organization may already provide useful information on the importance of individual nodes and connections, by identifying local features and investigating their importance for global network structure and function. As examples for biological networks I analyzed cortical fiber networks, metabolic networks as well as protein-protein interaction networks and, for comparison, also an artificial network, the German highway system.

5.2.2 Data and Methods

Analyzed networks

Brain networks.

I investigated long-range fiber projections in the cat and macaque brain. Nodes were brain regions or areas (for example, V1), and edges were fiber connections between them (cf. section 2.1.1). I considered connectivity data for a non-human primate, the Macaque monkey (73 nodes; 835 edges; density 16%), and the cat (55 nodes; 891 edges; density 30%). The clustering coefficient of these networks ranged between 40 to 50% (cf. Table 5.1).

Protein-protein interactions.

As an example of biochemical networks, I examined the protein-protein interactions of the *S. cerevisiae* yeast proteome (Jeong et al., 2001) (cf. section 2.1.3).

Metabolic networks.

Cellular metabolic networks of different species were analyzed. Metabolic substrates were nodes, and reactions were considered as edges (Ravasz et al., 2002) (Data at http://www.nd.edu/~networks/database/). For bacteria, I investigated the metabolism of *E. coli* with 765 metabolites and 3,904 reactions. For eukaryotes, the data for *Arabidopsis thaliana* (299 metabolites and 1,276 reactions) and the yeast *S. cerevisiae* (551 metabolites and 2,789 reactions) were examined.

Transportation network.

Data for the German highway (Autobahn) system were explored as an example for a transportation network (cf. section 2.2.1 for further information).

Benchmark networks.

Twenty random networks with 73 nodes and comparable density as the Macaque network were generated. Moreover, twenty scale-free networks with 73 nodes and equivalent density were grown by preferential attachment (Barabási & Albert, 1999), starting with an initial matrix of 10 nodes.

In addition to the random and scale-free networks, which consisted of only one cluster, networks with multiple clusters were considered. Twenty networks were generated with 72 nodes in order to yield three clusters of the same size with 24 nodes each. Connections within the clusters were distributed randomly, and six inter-cluster connections were defined to mutually connect all clusters. Average density of these networks was again similar to the Macaque data.

Prediction measures for edge vulnerability

I tested four candidate measures for predicting vulnerable edges in networks. All algorithms were programmed in Matlab (Release 12, MathWorks Inc., Natick) as well as implemented in C (C++ Builder 5.5.1, Borland Inc., Scotts Valley) for larger networks. Links were analyzed as directed connections for all networks.

First, the product of the degrees (PD) of adjacent nodes was calculated for each edge. A high PD indicates connections between two hubs which may represent potentially important network links.

Second, the absolute difference in the adjacent node degrees (DD) of all edges was inspected. A large degree difference signifies connections between hubs and more sparsely connected network regions which may be important for linking central with peripheral regions of a network.

Third, the matching index (MI) (Hilgetag et al., 2002) was calculated as the number of matching incoming and outgoing connections of the two nodes adjacent to an edge, divided by the total number of the nodes' connections (excluding direct connections between the nodes; Sporns, 2002). A low MI identifies connections between very dissimilar network nodes which might represent important 'short cuts' between remote components of the network.

Finally, edge frequency (EF), a measure similar to 'edge betweenness' (Girvan & Newman, 2002; Holme et al., 2002), indicates how many times a particular edge appears in all pairs shortest paths of the network. This measure focuses on connections that may have an impact on the characteristic path length by their presence in many individual shortest paths. I used a modified version of Floyd's algorithm (Cormen et al., 2001) to determine the set of all shortest paths and calculate the frequency of each edge in it. Multiple shortest paths between nodes i and j were present in the analyzed data sets. However, the standard algorithm only takes into account the first shortest path found. In order to account for edges in alternative same-length shortest paths, the EF was calculated as the average of 50 node permutations in Floyd's algorithm. This led to an increased prediction correctness of this measure in all networks; however, correlations between prediction and actual damage already converged after 10 permutations.

The component of the algorithm calculating the random permutation and the shortest paths is shown here as pseudo-code:

```
r = randperm(n);     % random permutation of node order
for k = 1:n
```

```
    for i = 1:n

     for j = 1:n

      if (path_length(i -> r(k)) + path_length(r(k) -> j) < path_length(i -> j)

        path_length(i -> j) = path_length(i -> r(k)) + path_length(r(k) -> j)

        path(i -> j) = path(i -> r(k) -> j);

      end; % if

     end; % for j

    end; % for i

end; % for k
```

Another possible prediction measure, not used here, would be the range of an edge (Watts, 1999; Sporns, 2002), that is, the length of the shortest path between two adjacent nodes after the edge between them is removed. For dense networks, such as cortical connectivity, only range values of 2 and 3 occurred. Having only two classes of range values was not sufficient to distinguish vulnerable edges in detail. However, the range may be a useful predictor for sparse networks with higher ASP.

5.2.3 Results

The elimination of an edge from a network can have two possible effects on the ASP. First, the parts previously connected by this edge can still be reached by alternative pathways. If these are longer, the ASP will increase. Second, the eliminated edge may be a *cut-edge*, which means that its elimination will fragment the network into two disconnected components. The probability for fragmentation, naturally, is larger in sparse networks. Network separation causes severe damage, as interactions between the previously connected parts of the network are no longer possible. Therefore, this impact can be seen as more devastating than the first effect, which may only impair the efficiency of network interactions. The applied ASP calculation disregarded paths between disconnected nodes, which would be assigned an infinite distance in graph theory. Therefore, the ASP in disconnected networks was actually shorter, because paths were measured within the smaller separate components. Cut-edges, which lead to network fragmentation, frequently occurred in the highway network (30% of all edges) and the yeast protein-protein interaction network (23% of all edges). However, cut-edges did not occur for cortical networks of cat and macaque and only to a limited extent (<5% of all edges) in the studied metabolic networks.

In the present calculation both increase and decrease of ASP indicate an impairment of the network structure, I therefore took the deviation from the ASP of the intact network as a measure for damage. I evaluated the correlation between the size of the prediction measures and the damage (Tab. 5.1 for all networks). While most of the local measures exhibited good correlation with the impact on ASP in real-world networks, the highest correlation was consistently reached by the EF measure. For the cortical networks, the measures of matching index and difference of degrees also show a high correlation.

Cortical connectivity differed from the other networks not only by the performance of different edge vulnerability predictors, but also in the density of connections and the amount of clustering. The cortical networks showed a higher density than the biochemical metabolic or protein-protein interaction networks. Whereas

Table 5.1: Density, clustering coefficient (CC), average shortest path (ASP) and correlation coefficients r for different vulnerability predictors of the analyzed networks (the index refers to the number of nodes). Tested prediction measures were the product of degrees (PD), absolute difference of degrees (DD), matching index (MI), and edge frequency (EF).

	Density	CC	ASP	r_{PD}	r_{DD}	r_{MI}	r_{EF}
Macaque$_{73}$	0.16	0.46	2.2	0.10**	0.57**	-0.40**	0.84**
Cat$_{55}$	0.30	0.55	1.8	0.08*	0.48**	-0.34**	0.77**
AT$_{299}$ (metabolic)	0.014	0.16	3.5	0.04	0.09**	-0.11**	0.74**
EC$_{765}$ (metabolic)	0.0067	0.17	3.2	0.31**	0.38**	-0.15**	0.75**
SC$_{551}$ (metabolic)	0.0092	0.18	3.3	0.11**	0.22**	-0.04	0.74**
SC$_{1846}$ (protein interactions)	0.0013	0.068	6.8	0.24**	0.02	-0.14**	0.60**
German highway$_{1168}$	0.0018	0.0012	19.4	0.19**	0.06**	-0.04	0.63**
Random$_{73}$	0.16	0.16	1.7	0.02	0.06	0.00	0.03
Scale-free$_{73}$	0.16	0.29	2.0	0.03	0.08	-0.01	0.03

* Significant Pearson Correlation, 2-tailed 0.05 level.

** Significant Pearson Correlation, 2-tailed 0.01 level.

the highway network showed similar density to the biochemical networks, its clustering coefficient was much lower because of the high proportion of linear paths in the highway network. The random and scale-free benchmark networks—designed to resemble size and edge density of the macaque cortical network—are presented at the end of the table.

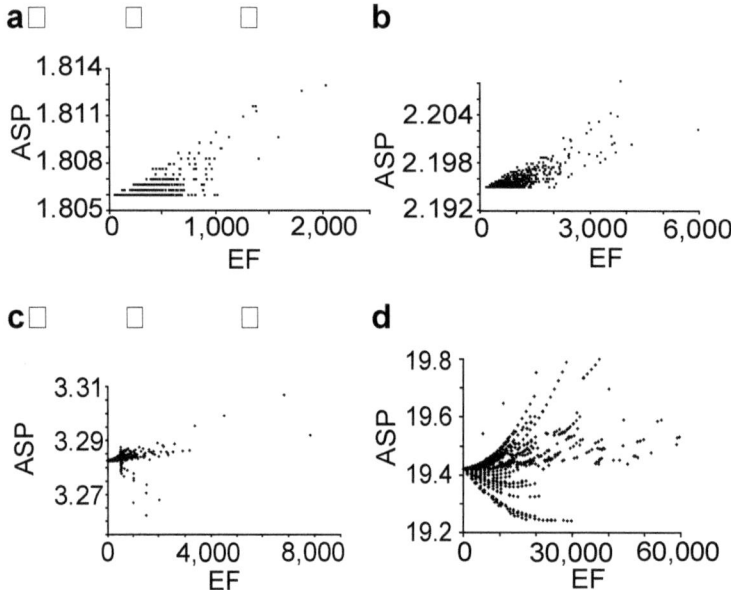

Figure 5.1: Frequency of edges in the all-pairs shortest paths and resulting network damage after elimination. **a**, Cat brain connectivity and primate (Macaque) brain connectivity **b** show a strong correlation with damage. **c**, Metabolic network of *S. cerevisiae*. **d**, German highway system. Decreases of ASP were caused by eliminated cut-edges, leading to a separation of the network.

Fig. 5.1 shows the ASP after edge elimination plotted against edge frequency (EF). For cortical networks (Fig. 5.1a,b) no network fragmentation occurred, and only an increase in ASP became apparent. For metabolic networks, for example, the network of *S. cerevisiae* (Fig. 5.1c), also a few cut-edges, lowering the ASP, were targeted. For the highway network containing linear chains of nodes (Fig. 5.1d), many cut-edges were observed. The elimination of these links, therefore, resulted in two disconnected compartments, each of which had shorter path lengths.

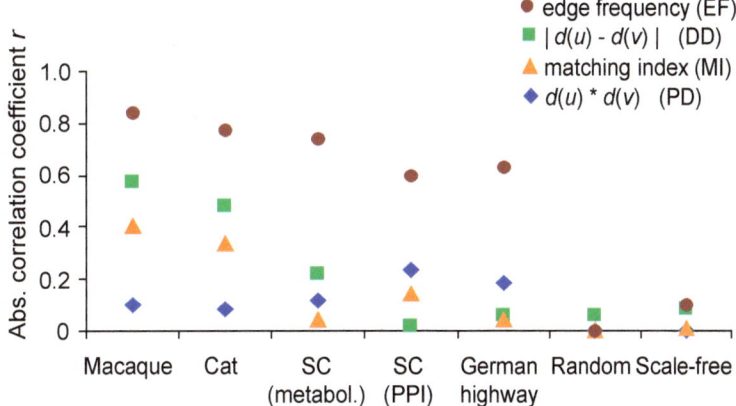

Figure 5.2: Evaluation of the performance of four predictors for edge vulnerability. Note that the absolute correlation coefficient was used (MI would have had negative r). For all networks, except the random and scale-free benchmark networks, edge frequency had the highest correlation with edge vulnerability. In random networks and scale-free networks with only one cluster, however, the tested measures were unable to indicate impact of edge elimination.

Comparison with benchmark networks

I also calculated the four predictive indices for scale-free and random benchmark networks with 73 nodes and a similar number of edges as the macaque cortical network. The comparisons focused on this network, because it showed the highest correlation between prediction measures and actual damage for all four measures. For the benchmark networks, however, all measures were poor predictors of network damage (Fig. 5.2). This is surprising, because scale-free networks generated here by growth and preferential attachment appeared to differ in their structure from real-world scale-free networks. Analyzing data for one of these real networks, the Internet at the autonomous systems level, which was previously shown to be scale-free (Barabási & Albert, 1999), I found that EF as a prediction measure was also performing well in this case ($r = 0.62$, not shown). The difference between the real and simulated scale-free networks may result from the fact that scale-free networks generated by growth and preferential attachment did not possess multiple clusters. I therefore tested whether the lack of connections between clusters might be the reason for the low performance of EF in the scale-free benchmark networks. I generated further 20 test networks; each consisting of three randomly wired clusters and six fixed inter-cluster connections (Fig. 5.3a). The inter-cluster connections (light gray) occurred in many shortest paths (Fig. 5.3b) leading to an assignment of the highest EF value, as no alternative paths of the same length were available. Furthermore, their elimination resulted in the greatest network damage as shown by increased ASP.

Network patterns in biological networks

After establishing the high impact on ASP of edges with large EF, I investigated what made specific edges more vulnerable than others. I here discuss three patterns that occurred in many of the analyzed networks. First, linear chains of nodes that appeared in biological as well as the artificial (highway) networks. Second, cycles of nodes that are also connected with the rest of the network and which particularly appear for metabolic networks. Third, clusters of highly interconnected regions of the network that occurred for all small-world networks, such as the analyzed cortical and biochemical systems.

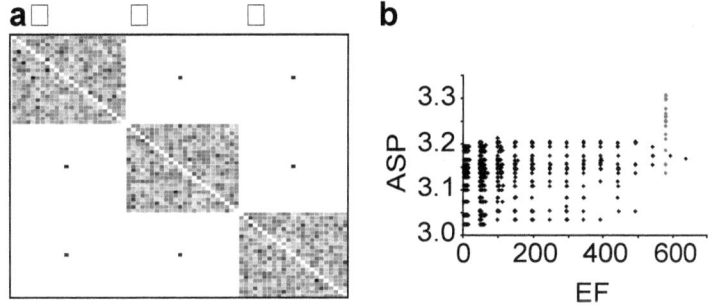

Figure 5.3: Connectivity for multi-clustered benchmark networks with comparable density to primate brain connectivity (cf. 5.2.2). The gray-level of a connection in the adjacency matrix indicates relative frequency of an edge in 20 generated networks. White entries stand for edges absent in all networks whereas black entries signify edges that were present in all networks. **a**, Connectivity of test networks with three clusters and six pre-defined inter-cluster connections. **b**, Edge frequencies in the all-pairs shortest paths against ASP after elimination of edges. Red data points represent the values for the inter-cluster connections in all 20 test networks. Inter-cluster connections not only have the largest edge frequency, but also cause most damage after elimination.

Naturally, other and more complex network patterns are possible. I already discussed elimination of edges between a hub and a node with fewer connections (cf. 5.2.4). Also, the functional role of these patterns (for example, feedback loops) was not examined here and would merit further study.

Linear chains

Linear chains of nodes with a terminal end (Fig. 5.4a) became apparent in various biological as well as artificial networks. These patterns were detected by testing for each node if it was part of a chain, that is, if it possessed exactly two undirected edges. In this case the chain was followed in both directions, and considered terminal, if at least one end of the chain had a terminal node. Using this method for identification, each terminal chain consisted of at least two edges. Nodes in the terminal chain were excluded from the further searching process. Terminal chains occurred frequently in the highway system, but also arose in metabolic networks in the form of redox chain reactions. For the highway system, the average terminal chain length was 6.4 edges, with a maximum of 22 edges. For the yeast protein-protein interaction network, 13% of the nodes were part of terminal chains, which were on average 2.25 edges long (maximum 6 edges). Eliminating edges at the terminal end of a chain would have a small impact, as only few nodes become disconnected from the rest of the network. On the other hand, severing the first edge that connects a chain to the rest of the network eliminates all paths leading to the chain nodes from the shortest paths matrix. The effect of eliminating edges in a chain can be seen clearly for the highway network (Fig. 5.1d). Edges that connect chains to the rest of the network have a large EF, and their elimination greatly decreases ASP, in contrast to edges at the terminal end. Indeed, for networks that show many cut-edges, also many terminal chains occurred. For the highway system, 42% of all nodes were part of terminal chains. Similar properties of edge vulnerability arise when the terminal of a chain end is formed by a small sub-network that is still smaller than the main network component at the start of the chain.

Cycles

Removal of edges near the connecting node will cause the largest increase in ASP, because all paths from the rest of the network have to pass most nodes of the cycle

(Fig. 5.4b). For edges 'opposite' the connecting node, however, only the paths for cycle nodes increase whereas the shortest paths from the huge rest of the network remain the same. Similar considerations apply if the cycle is connected to the rest of the network over multiple nodes.

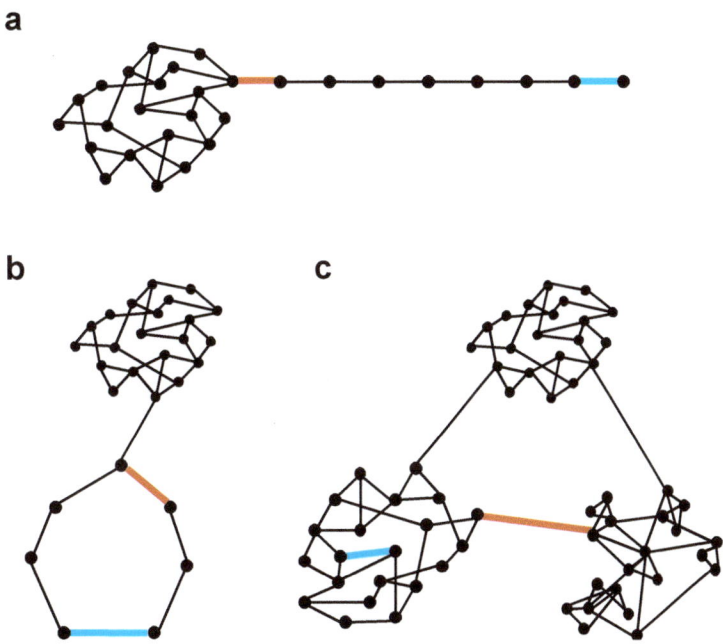

Figure 5.4: Network patterns and corresponding edge vulnerability. **a**, Elimination of edges of *linear chains of nodes* results in two disconnected components and a lower ASP. Edges eliminated at the proximal end of the chain (red line) cause a larger change in ASP than at the terminal end (blue line). **b** For *cycles* connected to the rest of the network with one node, edges near the connecting node (red line) are more vulnerable. Edges distant from this node (blue line) are less vulnerable, as their elimination only increases the shortest paths within the cycle, but does not affect shortest paths with the rest of the network. **c**, For *clustered networks*, edges within the clusters (blue line) can be replaced by several alternative pathways. Therefore, their elimination causes a smaller increase of ASP than that of edges between clusters (red line).

Clustered architecture

A clustered or modular architecture is a characteristic feature of many naturally occurring networks, such as cortical connectivity networks in the primate (Young, 1992; Young, 1993; Hilgetag et al., 2000) or the cat brain (Scannell et al., 1995; Scannell et al., 1999; Hilgetag et al., 2000) as well as metabolic networks (Ravasz et al., 2002). These systems are known to consist of several distinct, linked clusters with a higher frequency of connections within than between the clusters. Inter-cluster connections have also been considered important in the context of social contact networks, as 'weak ties' between individuals (Granovetter, 1973) and separators of communities (Girvan & Newman, 2002). I, therefore, speculated that connections between clusters might be generally important for predicting vulnerability (Fig. 5.4c). Whereas many alternative pathways exist for edges within clusters, the alternative pathways for edges between clusters may be considerably longer. Interestingly, previously suggested growth mechanisms for scale-free networks, such as preferential attachment (Barabási & Albert, 1999), or strategies for generating hierarchical networks (Barabási et al., 2001) did not produce distributed, interlinked clusters. Consequently, the low predictive value of EF in the scale-free benchmark networks was attributable to the fact that scale-free networks grown by preferential attachment consisted of one central cluster, but did not possess a multi-cluster organization and therefore had a lower diversity of edge vulnerabilities. This suggests that alternative developmental models may be required to reproduce the specific organization of biological networks, for example, spatial growth that can generate such distributed cluster systems (cf. chapter 4).

5.2.4 Discussion

I analyzed four measures for identifying vulnerable edges and predicting the impact of edge removal on global network integrity. Among these measures, the index of EF appeared consistently as the best predictor for damage to edges. The high performance of this measure may be linked to characteristic features of biological networks, as detailed below. For the macaque monkey, 7 out of the top 10 connections with highest edge frequency originated from, or projected to, the amygdala. In addition, these edges were most vulnerable as could be observed by the damage after edge elimination. Therefore, the amygdala appears to serve as a central link between many clusters of the network.

Following EF in terms of performance, the index for the difference of degrees also showed a high correlation in both the cortical and metabolic networks (Fig. 5.2). This means that connections between highly and sparsely connected nodes are vulnerable, especially in cortical networks. When a node with few connections is connected to an already well-connected node (hub), it can access a large part of the network, by using routes involving the hub. After eliminating the connecting edge, the node would have to use longer alternative pathways to reach the same parts of the network. This effect is particularly strong, if the node was the only one in its local neighborhood that was connected to the hub.

The matching index showed a large (negative) correlation with edge vulnerability in cortical networks and—to a lower degree—the protein-protein interaction network. Therefore, edges between dissimilar connections, that is, with low MI, were more vulnerable. This was due to the cluster structure of these networks.

Nodes with similar connectivity belong to the same cluster, and therefore multiple alternative pathways are available. Dissimilar nodes are more likely to be part of different clusters with few alternative pathways.

From theoretical studies it has been proposed for scale-free networks that edges between hubs are most vulnerable (Holme et al., 2002). However, in the networks analyzed here, both in the scale-free yeast protein-protein interaction network as well as for cortical networks, edges which connected nodes possessing many connections (large product of degrees) were not particularly vulnerable. Although a low increase can be seen for the correlation coefficient, edges with maximum vulnerability occurred for a small product of degrees.

One of the general advantages of using prediction measures, instead of testing the damage for all edges of a network, is computational efficiency. For a network with e edges and n nodes, the order of time for the calculation of EF that can predict the effect of edge elimination for all edges is $O(n^3)$. This is lower than for testing the damage after edge elimination for all edges individually, calculating the ASP e times resulting in a time complexity of $O(e \cdot n^3)$. The prediction measures presented here might therefore be particularly useful for large networks in which global testing is computationally impractical, or for networks with frequently changing connections demanding a regular re-calculation. Examples of such networks include acquaintance networks, Internet router tables, and traffic networks.

In the analyzed biological networks, inter-cluster projections may play an important role in linking functional units. For cortico-cortical networks, they connect and integrate different sensory modalities (for example, visual, auditory) or functional sub-components (Hilgetag et al., 2000). A lesion affecting these connections may result in dissociation disorders (Geschwind, 1965).

For metabolic networks, reactions proceed more frequently within a reaction compartment (for example, mitochondria and endoplasmatic reticulum) than between compartments. Therefore, in these systems as well, localized clusters arise, with many reactions within a compartment and few connections between compartments. Such an organization is also found in the investigation of protein-protein interactions and their spatial and functional clustering, in which fewer proteins from different groups interact (Schwikowski et al., 2000). Once again, interactions between proteins from different compartments correspond to inter-cluster connections, and might thus be among the most vulnerable edges of the network.

To a lower extent, 'inter-cluster' connections also appeared in the highway network, as the highway sub-systems of Western and Eastern Germany formed (spatially separate) dense regions connected by only four highways. That is, the elimination of four edges would once again split the German highway system into Eastern and Western components. The analysis of artificial networks was restricted to networks without functional differentiation of edges or nodes. It remains to be seen whether in functional or social networks, for instance, interactions of people with different functions within a company, may show a higher similarity in cluster-architecture with biological networks.

Clustered network architecture appears to result in edge-robustness in a similar way as scale-free architecture results in node-robustness. Random elimination of edges will most frequently select edges within a cluster. For paths routed through these edges, various alternative pathways exist, and the damage after edge elimination is small. Targeted attacks on inter-cluster connections, on the other hand, result in large network damage. I note that the cortical networks investigated here exhibit properties of both small-world (Hilgetag et al., 2000) and scale-free (Kaiser et al., 2005) networks and are therefore particularly robust to random failure of edges and nodes.

5.3 Effect of sequential node or edge removal

Scale-free networks have higher robustness than random ones against randomly located damage, whilst being sensitive to damage targeted at their most widely connected nodes (Barabási & Albert, 1999; Young et al., 2000) whereas randomly connected network do not display this increased sensitivity. In this section, I investigate whether cortical networks show similar properties.

5.3.1 Data and Methods

Cortical and benchmark networks

I used cortical connectivity networks of the cat with 65 areas and the macaque with 73 areas (cf. section 2.1.1 for details).

In addition, I constructed rewired, scale-free, random, and small-world networks to match the statistical properties (number of nodes and edges as well as similar clustering coefficient and ASP where possible) of the corresponding two brain structure networks. The networks were the same as used in section 3.2.2. Note that the algorithm for generating scale-free benchmark networks differed from the Barabási-Albert model, as described in that section, in order to match the statistical properties of the cortical networks.

Target determination

In order to determine the importance of a node to the overall network structure, a simple metric has been used, namely the number of connections formed by this node. In simulations requiring the targeted removal of nodes from the networks, the most highly connected node was eliminated.

To provide the corresponding metric for the targeted elimination of connections (edges) from the network, I chose edge frequency (similar to edge betweenness; Girvan & Newman, 2002), that is, the number of shortest paths between all pairs of nodes that pass though the edge. Edges with high edge frequency were chosen for targeted attack. Indeed, edge frequency had been shown to highly correlate with structural network damage for cortical as well as other biological networks (Kaiser & Hilgetag, 2004a, see also the previous section).

Network characterisation

The clustering coefficient shows the fragmentation of the network. The coefficient itself is a local property of each node and the average coefficient of all nodes is determined to be the clustering coefficient of the graph. This is a measure of how well connected the nodes of the network are.

Following Barabási and Albert (1999), I considered the average shortest path

(ASP) to characterize the network connectivity and integrity. The ASP between any two nodes in the network is the number of sequential connections that are necessary, on average, to link one node to another by the shortest possible route (Diestel, 1997). In case a network becomes disconnected in the process of removing edges/nodes and there is no path between two nodes, the pair of nodes is ignored. If no two connected points are left, the average shortest path is set to zero. I used Floyd's algorithm to determine the matrix of the shortest paths between each pair of nodes (Cormen et al., 2001). Note that due to the directed edges, the shortest path from node i to node j may not be of the same length as that from node j to node i.

Analysis methods

I used iterative random and targeted removal of nodes and connections to analyze the robustness of the networks against damage. Random removal means that a node or connection was selected randomly and deleted from the graph. In the case of targeted removal, the most important node or connection left in the network was selected (see above). After each deletion, the ASP of the resulting graph was calculated. The removal of nodes or connections continued until all nodes were removed from the network. To derive estimates of the variability in these connectivity measures, I considered 50 benchmark networks for each condition. In the cases of random removal, I repeated the analysis for the brain networks 50 times as well.

5.3.2 Results

Sequential elimination of nodes

I tested the influence of sequential node elimination on the network structure. Nodes were removed one by one from the network, either randomly or targeted. Plotting the ASP as a function of the fraction of deleted nodes illustrates the characteristic structural disintegration of each network type.

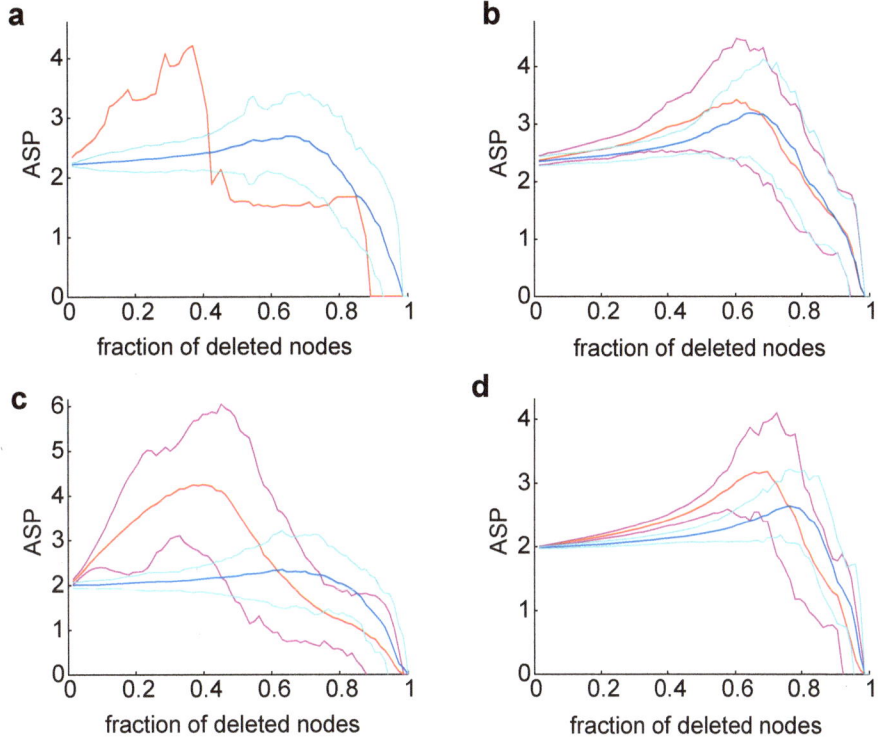

Figure 5.5: Sequential node eliminations in Macaque cortical as well as benchmark networks. The fraction of deleted nodes (zero for the intact network) is plotted against the average shortest path (ASP) after node removals. Nodes were removed either randomly (blue curve with light-blue indication of standard deviation) or taking out in order of connectivity starting with the most highly connected nodes (red curve with magenta lines for standard deviation). **a**, Cortical network. **b**, Small-world benchmark network. **c**, Scale-free benchmark network. **d**, Random benchmark network.

Fig. 5.5a illustrates the effect of random and targeted removal of nodes from the Macaque brain network. Clearly, the specific decline in ASP is different for the two analysis strategies. Whilst the random removal causes only a slow rise in the ASP, targeted removal of highly connected nodes has a much stronger effect on the network structure. Panels b - c contrast this specific curve to those observed when removing nodes from the different benchmark networks in a targeted fashion.

Whereas the ASP in the random and small-world networks is hardly affected by the targeted elimination of a large proportion of nodes, in the scale-free, like in the brain networks, the effect of targeted node elimination manifests itself in a sharp rise in this measure. Moreover, both, the scale-free and the brain networks show a decline in the ASP around the fraction of deletions, and the characteristic behavior of the brain network is mostly within the range encountered for the set of scale-free networks used. This is not the case for the other benchmark networks considered (see Fig. 5.5).

The decline in ASP at a later stage during the elimination process, as observed for the brain and scale-free networks, may appear unusual and deserves some additional attention. It can have two reasons: First, it could be that the network gets fragmented into different disconnected components. Each of these is smaller, and likely to have a shorter ASP. Second, the overall decrease in network size with successive eliminations could lead to a decrease in shortest path. This is, however, likely to be a slow process, as it will usually be offset by an increase in ASP due to the targeted nature of the elimination.

In order to quantitatively compare the different graphs, I considered two mea-

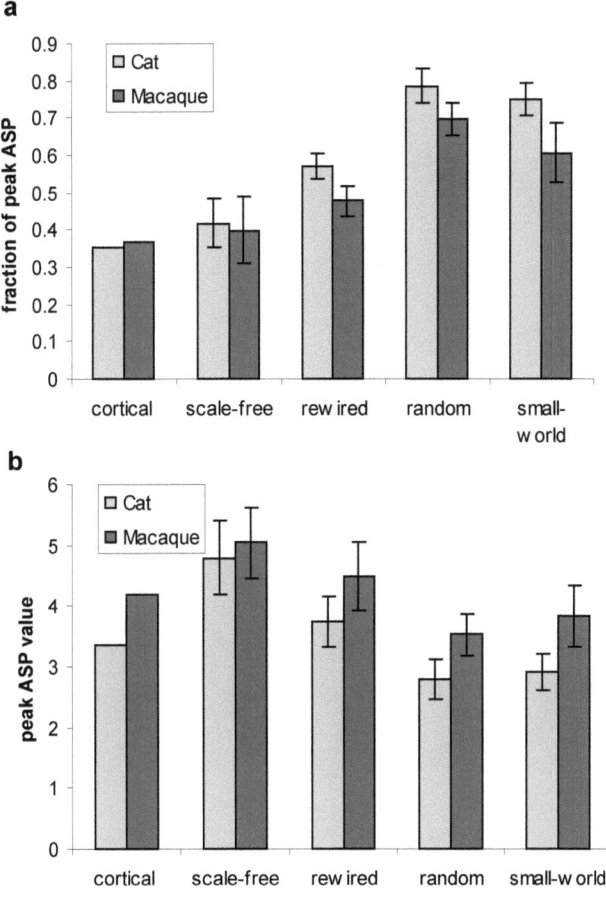

Figure 5.6: Fraction and value of peak ASP for attack node elimination. The average values and standard deviations are shown for the 50 generated benchmark networks. **a**, For the cat cortical network, only the fraction of peak ASP for the scale-free network is close to the cat network whereas the fractions of other benchmark networks are significantly higher. The same is the case for the macaque cortical network. **b**, However, the peak value of the ASP is higher for scale-free networks than in cortical networks in contrast to more similar values for the other benchmark networks.

sures. The first is the maximal ASP; the second is the fraction of deleted nodes, for which the maximal ASP value occurs (Fig. 5.6). For the fraction of such peak ASP, only the scale-free benchmark networks are close to the cortical fraction whereas all other benchmark networks show significantly higher fractions. This means that both in the cortical as well as the scale-free networks the removal of highly-connected nodes leads to a rapid increase of ASP so that the fraction of deleted nodes where the maximum ASP occurs is earlier than for other networks. However, the peak value for scale-free networks is greater than that for cortical networks.

Sequential elimination of connections

I also tested the similarity of sequential connection elimination curves. Connections were eliminated one after another from the network, either randomly or targeted, that means, taking out connections with high edge frequency first. Again, I compare the maximal ASP and the fraction of deleted nodes, for which the peak ASP occurs.

As can be seen in Figure 5.7, only the scale-free benchmark networks yield similar values for both the cat and macaque network whereas other networks only yield similar values for one of these cortical networks.

5.3.3 Discussion

I compared the effect that the removal of nodes and connections had on the ASP found in the brain connectivity networks and their benchmark counterparts.

Although the ASP has no immediate physiological or functional meaning, these measures can be used to determine the nature of the brain structure networks considered. Node elimination corresponds indirectly to inactivation or lesion of the corresponding brain areas, and from this perspective, can be interpreted in terms of the brain's robustness to regional damage. The elimination of connections

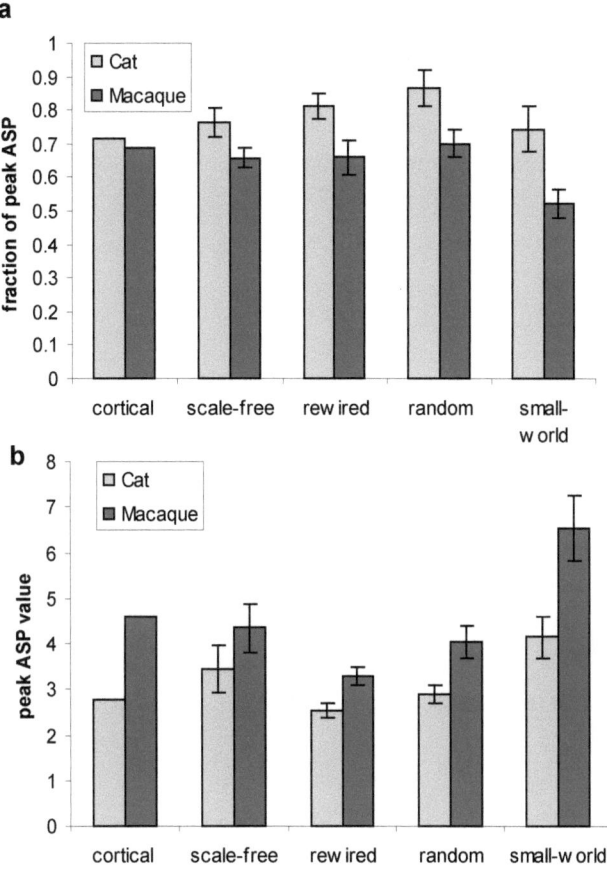

Figure 5.7: Fraction and value of peak ASP for attack connection elimination. The average values and standard deviations are shown for the 50 generated benchmark networks. **a**, For the cat network, scale-free and small-world fractions are similar to the cortical value whereas fractions of rewired and random networks are significantly higher. For the macaque network, however, all benchmark networks except for the small-world network show a similar fraction of peak ASP. **b**, The peak value of the cat cortical network can be matched by the random and rewired networks, nearly by the scale-free but significantly not by the small-world network. For the macaque, all networks except for the scale-free network show significantly different values.

corresponds indirectly to localized brain lesions that damage the white matter and interrupt communication between normally connected brain structures.

All these observations have been made equally during the analysis of the brain networks of cat and macaque, despite different edge densities in the two networks. It is therefore prudent to conclude that it may be extended to other mammalian brain networks. Hence, conditional robustness of brain function may be based to a large extent on two fairly simple structural properties of brain networks: the number of connections of individual nodes (Young et al., 2000), that means, their scale-free nature, and the heavily connected local clusters with fewer important 'bottlenecks' between them. Consequently, it appears feasible to determine the brain structures that are the most important to the maintenance of network function.

Is the brain optimized for high robustness or is robustness a by-product of other constraints? In my view, the emergence of highly-connected areas is more likely to be a side effect of brain evolution and development generating structures for efficient processing. For example, highly-connected areas (hubs) in the brain could play a functional role as integrators or spreaders of information (Sporns & Zwi, 2004). There are several potential developmental mechanisms for yielding brain networks with the highly-connected nodes typical for scale-free networks. Work in brain evolution suggests that when new functional structures are formed by specialization of parts of phylogenetically older structures, the new structures largely inherit the connectivity pattern of the parent structure (Preuss, 2000). This means that the patterns are repeated and small modifications are added during the evolutionary steps that can arise by duplication of existing areas (Krubitzer & Kahn, 2003). Such inheritance of connectivity by copying of modules is proposed to lead to scale-free metabolic systems (Ravasz et al., 2002). Alternatively, developmental algorithms based on the distance between areas have been proposed, and were shown to generate networks similar to cortical networks (Kaiser & Hilgetag, 2004c; Kaiser & Hilgetag, 2004a).

In conclusion, I have shown that cortical networks are affected in ways similar to scale-free networks following the elimination of nodes or connections. In the future, it would be interesting to compare the effect of experimental lesions with the simulated lesion approach used here. I therefore hope that this theoretical approach will prove useful in modeling robustness towards lesions.

5.4 Multiple lesions

5.4.1 Introduction

Traditional studies of biological systems have focused on the failure of single entities (structures or their links). In neuroscience, lesion studies eliminating one cortical area have been applied to various species and the resulting behavioral deficit has been observed. However, damaging multiple areas can lead to surprising results, for example, an enhancement of function compared to a single lesion for the paradoxical lesion effect in cats (Young et al., 1999; Hilgetag, 2000).

In gene-expression networks, the effect of gene knockout on the performance of the metabolism, and the lethality of gene damage, have been examined. However, the search for single genes that play a unique and significant role in diseases has been less successful than hoped for. Indeed, a large percentage of single gene knockouts appear to have very little effect on metabolism. For example, single gene knockout experiments for all genes on chromosome 5 of *S. cerevisiae* had little or no detectable effect on growth rate in 40% of the cases (Smith et al., 1997). Therefore, many deficits and molecular causes for diseases are believed to result from the accumulation of individually neutral mutations in multiple genes.

These examples suggest that studying the impact of multiple lesions is important for understanding the structure and function of biological and technical systems; yet few experimental and theoretical studies have worked on this problem. In this section, I investigate how multiple lesions (double and triple) deviate from the expected effect derived from single lesion experiments. Moreover, I apply the Multi-perturbation Shapley value Analysis (MSA) (Keinan et al., 2004) to asses the role of individual elements during the systems' performance in the intact and multiply lesioned state.

5.4.2 Data and Methods

Analyzed networks

I tested multiple lesions in the cortical networks of the cat (55 nodes, representing cortical areas) and the macaque monkey (73 nodes). In addition, I tested the metabolic networks of *S. cerevisiae* (551 nodes representing metabolites and 2,789 reactions / directed edges with 956 enzymes) and *E. coli.* (765 nodes and 3,904 reactions / directed edges with 1,506 enzymes/intermediate states) obtained from www.nd.edu/~alb.

Multiple lesions or enzyme removals

I tested the effect of removing multiple nodes of the network on the performance of the network (see below). Results from eliminating two or three nodes in the cortical networks were compared with predictions from the individual performances after single node removal. The predicted impact of double lesions of areas A and B was calculated using the performances P after single lesions of A or B and the performance of the intact network (baseline):

$$pred_{double}(A \wedge B) = P_{single}(A) + P_{single}(B) - P_{baseline}$$

Similarly, the predicted performance after triple lesions to some areas A, B and C involved the single lesion results reduced by two times the baseline:

$$pred_{triple}(A \wedge B \wedge C) = P_{single}(A) + P_{single}(B) + P_{single}(C) - 2 * P_{baseline}$$

In addition, I tested removal of multiple enzymes in metabolic networks. Because enzymes could be involved in several reactions, eliminating a single enzyme resulted in eliminating several reactions from the system, that means, removing multiple (typically 1-5) edges of the network. I looked at the effect of the removal of up to 10 enzymes with 1,000 combinations of enzymes for each number of removed enzymes. I observed how many paths, that means, pairs of nodes with an existing path between them, in the network were lost after the removal. Technically, this was achieved by evaluating the elements of the all-pairs shortest path matrix which give the shortest distance between each pair of nodes. The number of non-reachable pairs of nodes (paths with infinite distance) was compared before and after enzyme removal. If the removal did not result in additional lost paths, the removal was considered neutral, because as many metabolites as before could be reached within the reaction network. Looking at the neutral removals provides an estimate of how alternative pathways could compensate for the removal of reactions from the network.

Performance measures

I tested several performance measures to indicate how the network was affected by area or enzyme removal, respectively. Note, however, that the assessment of performance was based only on structural measurements and not on the dynamic behavior of the network.

Number of components: the number of disconnected fragments of the network. Although being conceptually simple, the measure has several disadvantages, such as that multiple components only occur after removing many network elements. Therefore, single and double lesions have usually no effect.

Average shortest path (ASP): average distance (number of edges) that has to be crossed to reach one node from another. The measure shows a visible effect for single and multiple lesions, however, it can decrease after network fragmentation so that it does not correlate linearly with network damage.

Clustering coefficient (CC): Percentage of direct neighbors of a node that are connected (Watts & Strogatz, 1998).

Clique number: A clique of a graph is its maximal complete subgraph. The problem of finding the size of a clique is NP-complete; however, a lower bound for the number of graph vertices in the largest clique $\omega(G)$ is given by $\omega(G) \geq \sum_i \frac{1}{n-d_i}$ where d_i is the degree of graph vertex i. In the remaining section, clique number refers to such lower bound for the actual value.

Multi-perturbation Shapley value Analysis (MSA)

MSA (from: http://www.cns.tau.ac.il/new/msa.html) is a framework for deducing causal function localization from multiple perturbations data. It provides insights to the workings of a given system, the "players" taking part in carrying out the different functions and the functional interactions between those "players". Essentially, it makes use of a set of multiple perturbations that are afflicted upon the system, while measuring the system's performance score in each, to quantify the importance of each of the elements, as well as the interactions between them.

Algorithms for network generation, calculation of network parameters and MSA used here were implemented in Matlab (Release 12, MathWorks Inc., Natick). The programs for MSA were developed by Alon Keinan, Alon Kaufmann and Dudi Deutscher at the Complex Network Systems' Lab, School of Computer Sciences, Tel-Aviv University.

5.4.3 Results

Prediction of the effect of multiple lesions

While there was in general a high correlation between the prediction based on the linear contribution of single lesions and the actual performance after double and triple lesions, some combinations showed much lower or higher performance values.

For the ASP as measure for the performance after elimination, the following pattern could be observed. After double lesions, the predicted ASP underestimated the actual ASP if both nodes were highly connected (hubs). Contrary, the prediction overestimated the ASP if one node was a hub and the other node was a sparsely connected node. In addition, overestimation also occurred for some combinations where both nodes were sparsely connected. This pattern occurred both for the cat and the macaque network.

Triple eliminations that differed from the linear prediction mostly involved nodes and pairs of nodes that already appeared for the double lesions. Interestingly, the top 10 of triple combinations differing from the predicted value for the cat and macaque cortical network only included one overestimated ASP (rank 6 in macaque). All other maximal deviations from predicted ASP showed underestimation and involved highly connected nodes, whereas the top 10 for double lesions showed overestimation in 30-50% of the ranks.

I further tested the clustering coefficient as performance measure for the macaque network (Fig. 5.8). In contrast to the ASP, no clear pattern for over- or underestimation relating to node degree occurred. However, the most highly-connected node in the macaque network (the Amygdala) was a member of 6 out of the 10 pairs with highest deviation from prediction in double lesions and occurred in all of the top 10 triple combinations. Until now, it is not clear whether this effect is due to the high degree of the node or is due to the position in the network (possibly linking different clusters). Overall in the top 10 of double as well as triple lesion effects, the linear prediction underestimated the real effect (increase of clustering coefficient) in 70% of the cases.

MSA for cortical networks

The results for the different prediction measures are described below:

Number of components (Fig. 5.9a): Significant contribution values of individual

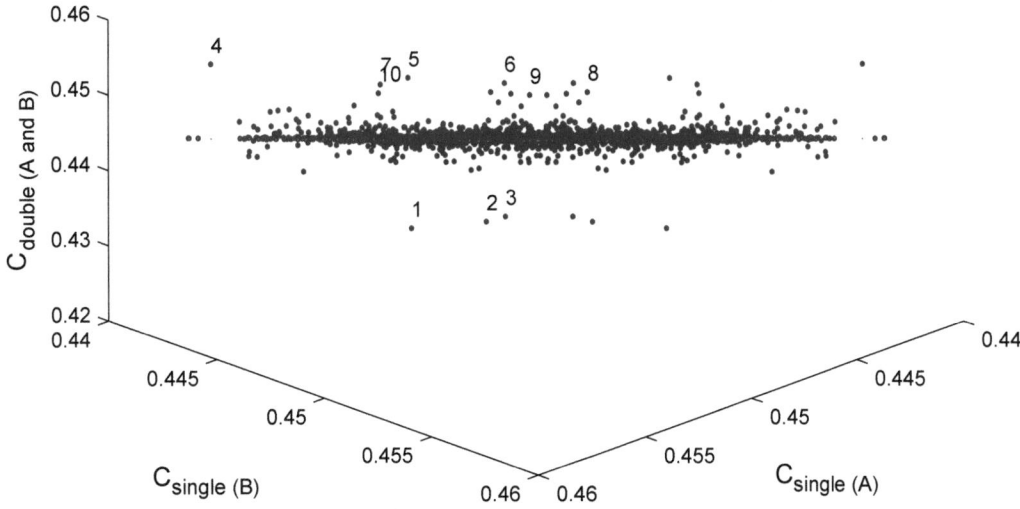

Figure 5.8: Macaque cortical network. Clustering coefficient after double lesions by removing nodes A and B (z-axis) plotted at the XY-coordinates given by single lesions of A or B. Most double lesions lie on a plane defined by the XY-coordinates (the plot shows a rotation of that plane in that deviation from the plane can be easily seen). However, some double lesions are outside the prediction plane (above or below). Numbers show the ranking in absolute deviation from the prediction (labels were only applied to one occurrence of that value, although each value existed at least twice as each pair of nodes led to calculations for the pairs (A,B) and (B,A)).

nodes only became apparent after more than half of the nodes were eliminated. The nodes with highest contribution to the number of disconnected components appeared to be nodes linking different clusters.

Average-Shortest Path (ASP) (Fig. 5.9b): Contributions of different nodes could be distinguished much earlier after about 20% of the nodes were removed. In addition, the contribution at the early stage seems to correlate with the final contribution, therefore removing up to 25% of nodes might be sufficient for an approximate but reliable analysis.

Clustering coefficient (CC) (Fig. 5.9c): In contrast, contributions to the clustering

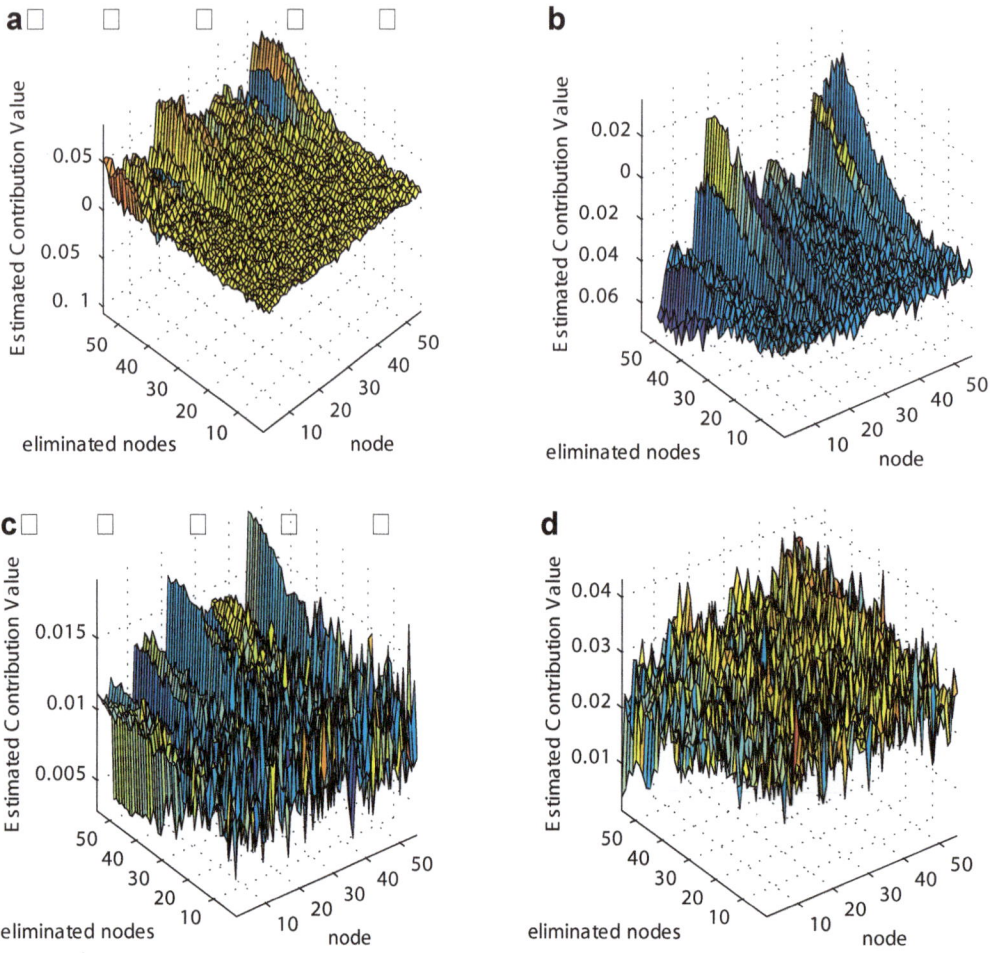

Figure 5.9: Multiple-Shapley-Analysis (MSA) for the cat cortical network. Estimated contribution values for different nodes depending on the number of eliminated nodes. The curves for different performance measures are shown. **a**, Number of (disconnected) components. **b**, Average-Shortest Path (ASP). **c**, Clustering coefficient. **d**, Clique number.

coefficient can only be assessed at a later stage of removal.

Clique number (Fig. 5.9d): There seems to be no pattern of contribution of nodes to the clique number. That means, there is homogeneous contribution to the approximated clique number.

Multiple enzyme removal in metabolic networks

The number of neutral knockouts, that means, enzyme elimination whose effects can be compensated by alternative pathways, is shown in Figure 5.10. Despite the fact that both the yeast and the *E. coli* metabolic networks are not very dense (density of 0.92% and 0.67% respectively), about 70% of single enzyme knockouts are neutral. In addition, the percentage of neutral removals decreases slowly. Even eliminating seven enzymes had no effect in 10% of the cases.

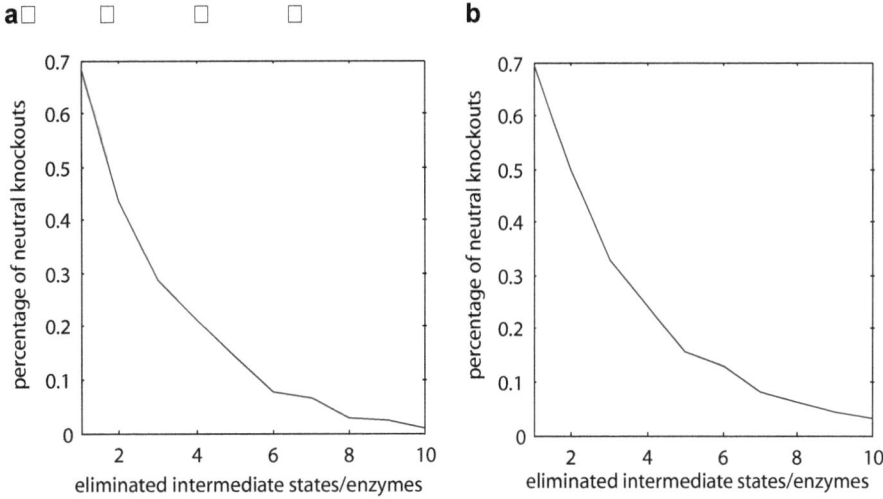

Figure 5.10: Ratio of neutral knockouts for multiple enzyme elimination (1.0=100%; 1,000 permutations for each number of eliminated enzymes). **a**, *S. cerevisiae* (551 metabolites; 2,789 reactions; density: 0.0092) **b**, *E. coli* (765 metabolites; 3,904 reactions; density: 0.0067)

5.4.4 Discussion

Multiple lesions of cortical networks

Certain double and triple lesions have a much larger or lower effect than predicted by the effect of single lesions. Such differences could explain surprising results in experimental lesion studies. Furthermore, the predictions for particular severe lesion configurations could be tested experimentally. In this way it can be examined to which extent experimental lesion deficiencies can be explained by the structural connectivity architecture.

MSA for cortical networks

'Inter-cluster' nodes at the border between clusters seem to have higher contribution values than nodes within a cluster. As cortical networks are dense, distinctly higher contribution values of some individual nodes only become apparent after more than half of the nodes are removed form the network. Of most relevance to neuroscience is the observation that single lesions are not sufficient to identify a nodes' contribution.

Multiple knockouts in metabolic networks

Single gene knockout studies of genes in yeast have shown that 80% of the genes do not appear to be essential for the survival of the organism. These can be due to duplicate genes, enzymes with similar structure (isoenzymes), or alternative pathways. The structural analysis concerning neutral knockouts yield an upper limit of 70% of single gene robustness that could be explained by alternative pathways. Naturally, the actual role of alternative pathways might be much lower as very long alternative pathways might be unfeasible for metabolic networks. In addition, it has been shown that the actual number of neutral gene knockouts is lower when the effect of knockouts is tested for all environmental conditions to which the organism can be exposed (Papp et al., 2004).

5.5 Summary

Biological systems can be remarkably robust towards damage. Studies in yeast have shown that in 80% of the cases single gene knockout have little effect on the fitness of the organism (Wagner, 2000) [1]. For the cortical networks, damage to cortical areas can in some cases be compensated by alternative intact areas. On the other hand, the elimination of some nodes or edges can lead to significant damage. Can this behavior be explained by network topology?

An analysis of the similarity between cortical networks—from the cat and the macaque monkey—and random, rewired, small-world, and scale-free benchmark networks shows that scale-free networks are most similar in several aspects. Most importantly, cortical as well as scale-free networks show a huge disparity between random elimination of nodes or edges and targeted elimination where the most vulnerable parts of the network are removed. Similarly, an analysis of edge vulnerability has shown that the targeted removal of connections between clusters has a large effect compared to the removal of connections within a cluster. This range of effects could correlate with the range of effects of lesions that were observed experimentally. However, a detailed comparison between the lesion effect observed *in vivo* and the effect predicted from network topology is needed in order to evaluate to which extend structure influences function.

Current analysis of network topology supports the notion that cortical systems are robust to random removal of areas or projections. A random lesion of nodes

---

[1] However, some of these genes might be important under different growth conditions (Papp et al., 2004).

will in most cases eliminate nodes with few connections which will have a low effect on the network. Similarly, random removal of edges will in most cases take out connection within clusters and rarely hit projections between clusters.

Simulations of multiple lesions can help to characterize the contribution of individual areas. In addition, simulated performances that were unexpected from the effect of single lesions were observed for double and triple lesions. Future studies will have to compare the simulated and the experimentally observed effect of lesions.

Chapter 6

General discussion and outlook

The cortical brain network of the cat or the macaque monkey as well as the neural network of the nematode *C. elegans* show an intricate organization. Both display clusters in their connectivity pattern representing functional groups in the cortex or ganglia in the nematode. In addition, both show features of small-world networks with low average-shortest path and relatively high clustering coefficient, as well as scale-free attributes. I have studied how these networks are organized concerning wiring lengths, how they—as well as other spatial networks—could have developed and how robust they are to damage.

6.1 Which constraints shape the brain?

Observing the spatial position of cortical areas, many long-distance connections were found, which seems counter-intuitive if saving wiring length is the only important constraint. Next, I investigated whether neural systems show component placement optimization. As has been stated in earlier studies "the best tests of the hypothesis will include at least a significant sample of a system at a given level of organization, not just a few isolated elements" (Cherniak, 1994). Therefore, I analyzed a large set of 95 macaque cortical areas as well as 256 neurons of *C. elegans*. In addition, I applied Euclidean metric distance instead of indirect measures used in earlier studies. For both networks, node arrangements with lower total wiring length are indeed possible. Therefore, static constraints such as the costs for establishment or maintenance of long-distance connections appear to compete with other constraints that make long-distance connections necessary.

These other constraints could be functional requirements for efficient processing. I could show that the number of intermediate processing units—as indicated by the average shortest path—is reduced by the existence of long-distance connections. This has three benefits for the organism. First, the introduction of time delays by additional nodes is prevented. The resulting faster reaction times can increase the evolutionary fitness of the organism. Using 'fight or flight' reactions quickly when the animal is threatened by a predator can reduce potential harm. Second, intermediate nodes could introduce 'noise' or unwanted processing to signals transmitted over cortical connections. For example, if only short-distance fibers would exist, but no direct connections between the visual cortex and the frontal cortex, information from the visual cortex would need to pass through the sensory-motor areas. The original visual information would interfere with other sensory modalities of information. Third, the existence of projections to neighboring as well as distant areas enables both target areas to receive and process information (almost) simultaneously. Synchronous processing was discussed as one possibility of binding different features of an object to a coherent perception of that object (von der Malsburg, 1981). For example, whereas shape and color of an object are processed in different parts of the brain, the features are integrated as belonging to one object.

It will be interesting to extend the current analysis by including different organizational levels. For example, will the same patterns occur for neural connectivity within cortical areas or are there different constraints that influence wiring organization?

In addition, there are still many open questions concerning the development of long-distance connections. For example, at which point during individual development are such connections established? Are the cues for axon growth to distant target areas different from cues for short-distance connections? How are specific long-distance connections such as the projections between V1 and area 46 in the macaque encoded? Or, alternatively, are there more long-distance connections during early development which are later removed again from the network?

6.2 How did the brain develop?

I have proposed a new method for generating spatial graphs, incorporating both limited and virtually unlimited growth, that can produce a variety of metric real-world networks. The metric is not limited to Euclidean space as in the discussed examples, but may also use measures of similarity to define the link probability, for example, social relations (Watts et al., 2002).

Spatial limits seem to occur in biological networks, as the skull as well as apoptosis factors limit neural growth at the network border and in the cell different reaction compartments (mitochondria, endoplasmatic reticulum, cell core, etc.) and distant positions of proteins in the membranes limit interaction. Furthermore, it has been shown, that the spatial separation of molecule interaction is critical in the early chemical evolution of metabolic pathways (Szabó et al., 2002).

Experimental studies also suggest that a mechanism without activity might be sufficient for the generation of the global cortical organization. It was shown experimentally that even without activity transmission along synapses, similar cortical connectivity and layer architecture can arise (Verhage et al., 2000). However, activity is needed again for the organization of cortical maps (for example, ocular dominance columns; Hubel et al., 1977).

Also, spatial growth yields networks with mostly short-distance connections. Again, such preference was found for neural networks both on the level of connections between neurons within an area (Braitenberg & Schüz, 1998) but also for the inter-area connectivity of cortical networks (cf. chapter 3).

Moreover, in man-made networks such as the Internet, and transportation networks the spatial distance limits edge formation and 'borders' may exist. For example, undersea connections are much more expansive (and therefore unlikely) than overland connections with the same distance. These geographic borders like sea, mountains, deserts and so on can limit network growth resulting in a higher clustering coefficient and density of the network and might be the reason for small-world properties.

In contrast to previously studied spatial graphs (Watts, 1999), networks generated by the model presented here were always connected. Moreover, the approach was able to generate small-world graphs, which is thought not to be possible in the spatial graph model in which positions are chosen randomly before edge formation (Watts, 1999). Finally, the model was also able to produce scale-free networks with relatively low maximum degree, similar to, for example, the German highway system or protein-protein interaction networks. A systematic evaluation of model parameter space was carried out at the specific network size of 100 nodes, which was feasible computationally. It would be interesting to evaluate larger or smaller network sizes and to investigate for them, whether small-world networks can be generated in a larger range of parameters α and β.

A model with three distinct classes of nodes and different time windows resulted in three clusters in the wiring architecture. Therefore, multiple clusters may arise from spatially distributed origins for spatial growth development as well as different time windows for establishment of connections.

Future research for cortical networks will have to incorporate a higher level of detail both in the analyzed network and in the proposed developmental algorithm. For instance, the modeled network should include information about in which layer projections arise or to which layer in the target area they project to. Similarly, nodes should be assigned information about their modality (for example, processing visual or auditory signals) or other functional properties. Furthermore, developmental algorithms should take into account different growth factors as well as the formation of the skull, vesicles, fiber bundles, or cortical areas that could influence axon growth.

6.3 Why is the brain robust?

The structure of cortical networks was found to facilitate robustness towards damage in several ways. First, the high neighborhood connectivity together with the existence of multiple clusters has an effect on the robustness towards edge removal (Kaiser & Hilgetag, 2004a). For connections within clusters, many alternative pathways of comparable length do exist once one edge is removed from the cluster. For edges between clusters, however, alternative pathways of comparable length are unavailable and removal of such edges should have a larger effect on the network.

Second, cortical as well as scale-free benchmark systems are robust to random node elimination but show a larger increase in ASP after removing highly connected nodes. Again, as for the edges, only few nodes are highly connected and therefore critical so that the probability to select them randomly is low.

In addition, multiple lesions in cortical networks were in some cases found to yield effects that could not be predicted from the combined effect of single lesions. Therefore, the MSA method might be necessary to better evaluate the contribution of individual cortical areas.

A design for...Robustness?

For random lesions, cortical networks were found to be robust in that elimination of edges or nodes had little effect on global network properties. Is the brain optimized for high robustness or is robustness a side effect of other (for example, processing) constraints? Although the question is difficult to answer right now, I want to discuss possible hints by examining functional reasons for underlying properties for high robustness such as having multiple clusters and highly connected network nodes.

In simple organisms like radial-symmetric Coelenterate, nerve cells form a nerve net with no spatial concentration of neurons. After the emergence of sensory or effector units, however, areas with higher densities of nerve cells (ganglia) arise. With increasing cephalization, a high concentration of nerve cells can be found in the head of the animal. Together with the establishment of functional units, more connections can be found within than between units. For the retina, there are more connections within the retina (every ganglion cell receives incoming connections from several bipolar and amacrine cells) than projections to the lateral geniculate nucleus (LGN). Similarly, at higher levels of visual processing in the macaque, less than 40% of excitatory synapses of a given volume of visual cortex come from other areas or thalamic nuclei, that means, intrinsic connectivity is higher (Young, 2000). Therefore, it is not surprising that also larger units—for example, the visual cortex—form clusters with more connections between visual areas than between visual areas and areas of other function or modality. Therefore, functional specialization might be a reason for cluster architecture.

Highly connected nodes could result from time windows during development (cf. section 4.4.3). For example, it has been reported in embryology that parts that are older phylogenetically, also develop earlier during individual (ontogenetic) development (Kahle, 1969; Hinrichsen, 1990). For the macaque brain connectivity network, I found that regions that develop earlier also have relatively more incoming connections. Of all connections between neocortex and archi- and paleocortex, 60% project to older areas (from neocortex to archi- and paleocortex) and only 40% in the reverse direction (Fig. 6.1). Within the neocortex, the projection pattern was more symmetric with 52% of the projections going from frontal, temporal and sensorimotor to parietal and occipital cortex vs. 48% in the other direction. In addition, the total number of connection seems to be higher for older nodes. This supports the idea that early areas can receive more incoming—and therefore more total—connections from other regions. This might be explained in that they can get incoming connections even after the time window for establishing own outgoing connections has passed.

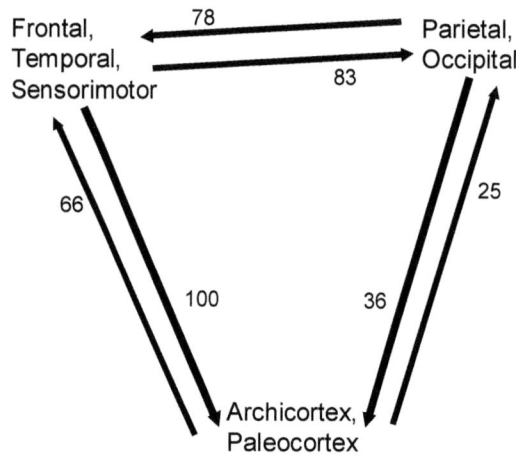

Figure 6.1: Projections between brain regions (Macaque data set including 73 nodes). Numbers denote the number of projections for each given direction (also indicated by the width of the arrows).

Up to this moment, I can only speculate on the function of the highly connected nodes. However, it seems that these nodes could be integrating information from their respective cluster and forward it to other clusters. For example, in the macaque brain network LIP (Lateral Intra-Parietal area) gets many incoming connections from visual areas of the parietal and occipital cortex and projects to the frontal cortex (FEF:Frontal Eye Field). The Amygdala receives incoming connections from many areas of the neocortex and thus might function as an integrating unit used for evaluating all aspects of external situations. Thus, integrating—possibly multi-modal—information and linking different clusters might be the role of highly connected nodes.

In general terms, robustness to lesions could just be a side effect of the plasticity of the cortex. Whereas robustness refers to the functional reorganization after internal change of removing connections or areas, plasticity refers to the ability to adapt to changes in general, that is, to learn or re-learn. In both cases, parts of the brain have to adapt to a largely different input pattern, and have to change synaptic properties and connections to ensure a behaviorally useful function. For an individual area, it might be impossible to distinguish whether changes in the input pattern are due to an altered environment of the organism or lesions within the brain. Therefore, it might be interesting to test experimentally, whether similar mechanisms are in place for dealing with internal and external changes.

However, robustness might be also important on an evolutionary scale. It was argued for metabolic systems that robustness towards gene deletion means a higher possibility for neutral mutations which might in turn increase genome variability. This could be helpful for the survival of a population in that individual members are able to adapt to changes in the environment (Wagner, 2005). Similarly, robustness of cortical networks might make 'experiments' with different connection configurations possible without having cognitive deficits in all cases. To my knowledge, there is currently not enough data available on connection variability and effects of such changes on functional performance in mammals.

In my opinion, biological networks at the level of cortical systems are not optimized for high robustness. Instead, robustness could be a side-effect of other functional constraints. For example, the formation of functional clusters is necessary for exclusion of signals with different modality and highly-connected areas could function as integrators of (multi-modal) information or spreaders of information to multiple clusters or many nodes within one cluster. Therefore, it will be interesting in the future to compare the function of areas with their number of projections and the function of their directly connected neighbors.

6.4 Possible applications

The results and proposed future directions of research may lead to several clinical, pharmaceutical, or industrial applications. Some of which are briefly outlined here:

- A theoretical framework for lesion effects and compensation can help to assess recovery potential. Eventually, it will be possible to "train" the region that will take over the function of damaged areas. For these applications, it is necessary to have human connectivity data. Currently discussed methods for yielding human connectivity data are diffusion tensor imaging (DTI) (Parker et al., 2002; Tuch et al., 2003) and functional connectivity (Bartels & Zeki, 2005; Eguíluz et al., 2005).

- Knowledge about robustness in neural systems could be compared with mechanisms for other biological networks, specifically to metabolic networks. The role of alternative pathways or protein similarities will not only help in assessing the function of healthy metabolic systems but also as a means to find critical components and therefore potential drug targets (Grigorov, 2005).

- Finally, one could implement features that cause high robustness of biological systems—unparalleled by technical systems—in computer architecture. This could complement a current trend in computer science towards recovery-oriented computing that shifts the emphasis from the prohibition of failures—which seems to be impossible in the end—to rapid recovery after failures (Patterson et al., 2002).

Glossary

Adjacency (connection) matrix: The adjacency matrix of a graph with n nodes is a $n \times n$ matrix with entries $a_{ij} = 1$ if node j connects to node i, and $a_{ij} = 0$ if there is no connection from node j to node i.

Anatomical connectivity: The set of physical or structural (synaptic) connections linking neuronal units at a given time. Anatomical connectivity data can range over multiple spatial scales, from local circuits to large-scale networks of inter-regional pathways. Anatomical connection patterns are relatively static at shorter time scales (seconds to minutes), but may be dynamic at longer time scales (hours to days), for example, during learning or development.

Characteristic path length (Watts & Strogatz, 1998): The characteristic path length L (also called "path length" or "average shortest path") is given by the global mean of the finite entries of the distance matrix. In some cases, the median or the harmonic mean may provide better estimates.

Clustering coefficient (Watts & Strogatz, 1998): The clustering coefficient C_i of a node i is calculated as the number of existing connections between the node's neighbors divided by all their possible connections. The clustering coefficient ranges between 0 and 1 and is typically averaged over all nodes of a graph to yield the graph's clustering coefficient C.

Component: A component is a set of nodes, for which every pair of nodes is joined by at least one path.

Connectedness: A connected graph has only one component. A disconnected graph has at least two components.

Degree: The degree of a node is the sum of its incoming (afferent) and outgoing (efferent) connections. The number of afferent and efferent connections is also called the in-degree and out-degree, respectively.

Density (Edge density): Proportion of edges (or arcs) existing in the network to the number of all possible edges (arcs).

Distance: The distance between a source node j and a target node i is equal to the length of the shortest path.

Distance matrix: The entries d_{ij} of the distance matrix correspond to the distance between node j and i. If no path exists, $d_{ij} = \infty$.

Effective connectivity: Describes the set of causal effects of one neural system over another (Friston, 1994; Sporns et al., 2004). Thus, unlike functional connectivity, effective connectivity is not "model-free", but requires the specification of a causal model including structural parameters. Experimentally, effective connectivity may be inferred through perturbations, or through the observation of the temporal ordering of neural events.

Exponential graph: Erdös-Renyi random graph (Erdös & Rényi, 1960) with binomial degree distribution that can be fitted by an exponential function. In this thesis, such a graph is referred to as *random network*.

Functional connectivity (Sporns et al., 2004): Captures patterns of deviations from statistical independence between distributed and often spatially remote neuronal units, measuring their correlation/covariance, spectral coherence or phase-locking. Functional connectivity is time-dependent (hundreds of milliseconds) and "model-free", that means, measuring statistical interdependence (mutual information), without explicit reference to causal effects.

Graph: Graphs are a set of n nodes (vertices, points, units) and k edges (connections, arcs). Graphs may be undirected (all connections are symmetrical) or directed. Because of the polarized nature of most neural connections, I focus on directed graphs, also called digraphs. In addition, graphs are simple, that means, multiple (undirected) edges between nodes or loops (connections of one node to itself) do not exist.

Hodology: The study of pathways. The word is used in several contexts. (1) In brain physiology, it is the study of the interconnections of brain cells. (2) In philosophy, it is the study of interconnected ideas. (3) In geography, it is the study of paths.

Linear graph: Graph with many linear chains of nodes which can be detected by the *clustering coefficient* being lower than the *density*.

Matching index: The matching index of two network nodes is defined as the average percentage of identical incoming or outgoing directed edges of the two nodes (Sporns, 2002).

Optimal component placement (also: Component Placement Optimization, CPO): Nodes of a spatial network are arranged in an optimal way, that means, every permutation of node positions (with connections being unchanged) would lead to a higher total wiring length (Cherniak, 1994).

Path: A path is an ordered sequence of distinct connections and nodes, linking a source node j to a target node i. No connection or node is visited twice in a given path. The length of a path is equal to the number of distinct connections.

Random network: A graph with uniform connection probabilities and a binomial degree distribution (Erdös & Rényi, 1960). All nodes have roughly the same degree ('single-scale').

Scale-free graph: Graph with a power-law degree distribution. 'Scale-free' means that degrees are not grouped around one characteristic average degree (scale), but can spread over a very wide range of values, spanning several orders of magnitude.

Small-world graph: A graph where the *clustering coefficient* is much higher than in a comparable random network but the *characteristic path length* remains about the same. The term 'small-world' was coined by the notion that any two persons can be linked over few intermediate acquaintances (Milgram, 1967).

Spatial graph: Graphs or networks that extent in space, that means, that every node has a spatial position. Spatial graphs are usually analyzed as being two- or three-dimensional but naturally more dimensions are possible (Watts, 1999).

Spatial growth: A new method to yield *spatial graphs* (Kaiser & Hilgetag, 2004c). Starting with one node, at each step a new node with a random spatial position is added to the network. Then, the probability of the new node to establish connections with existing nodes decays with the spatial distance (for example, exponential decay). Thereby, connections to nearby nodes are more likely than to distant nodes. If a node does not establish any connections, it is removed from the network. The procedure is repeated until the desired number of nodes is generated.

Cortical connectivity matrices

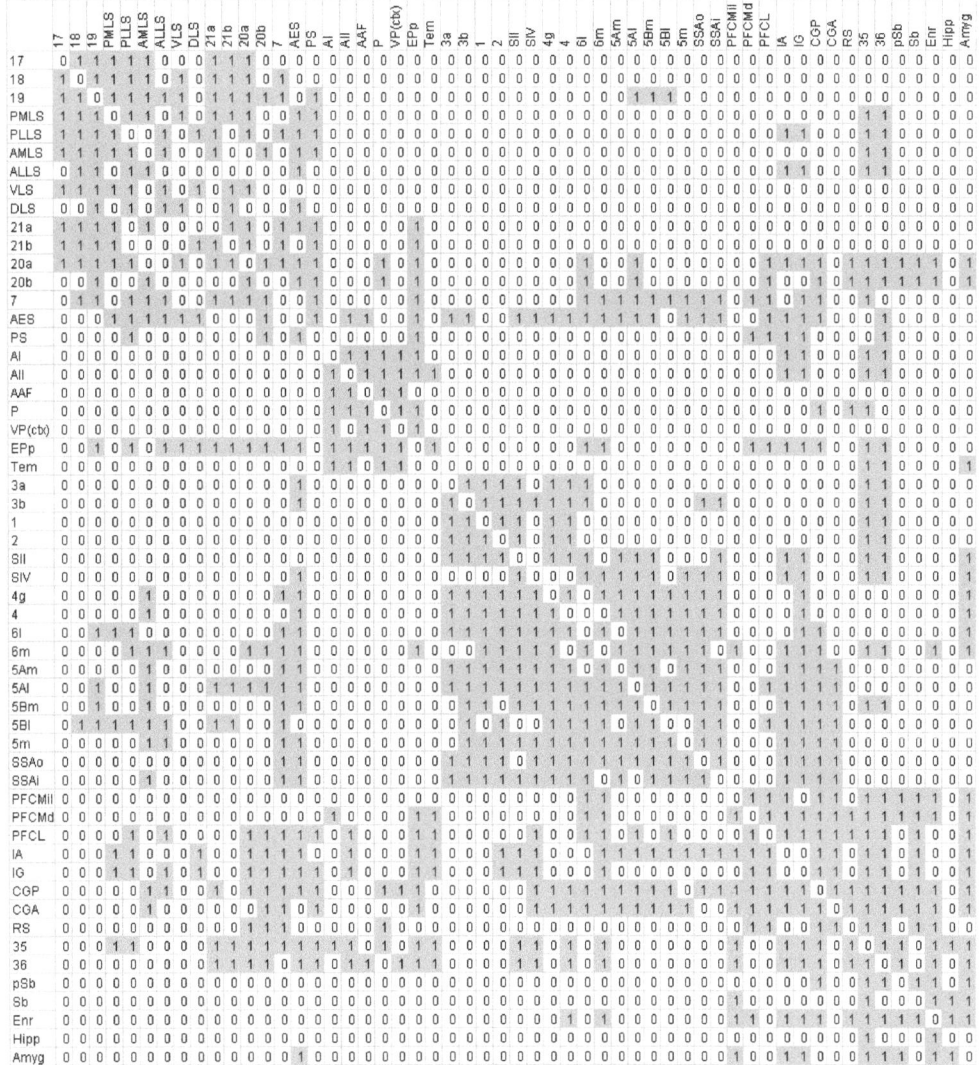

Figure A.1: Cat connectivity matrix of 55 cortical regions and 891 directed connections. A zero entry stands for connections that were either found absent or for which no search was done. Existing connections are assigned the value one (and shown with gray background for easier visibility). Rows represent outgoing and columns incoming connections of a region.

Figure A.2: Cat connectivity matrix of 65 cortical regions and 1,139 directed connections. A zero entry stands for connections that were either found absent or for which no search was done. Existing connections are assigned the value one (and shown with gray background for easier visibility). Rows represent outgoing and columns incoming connections of a region.

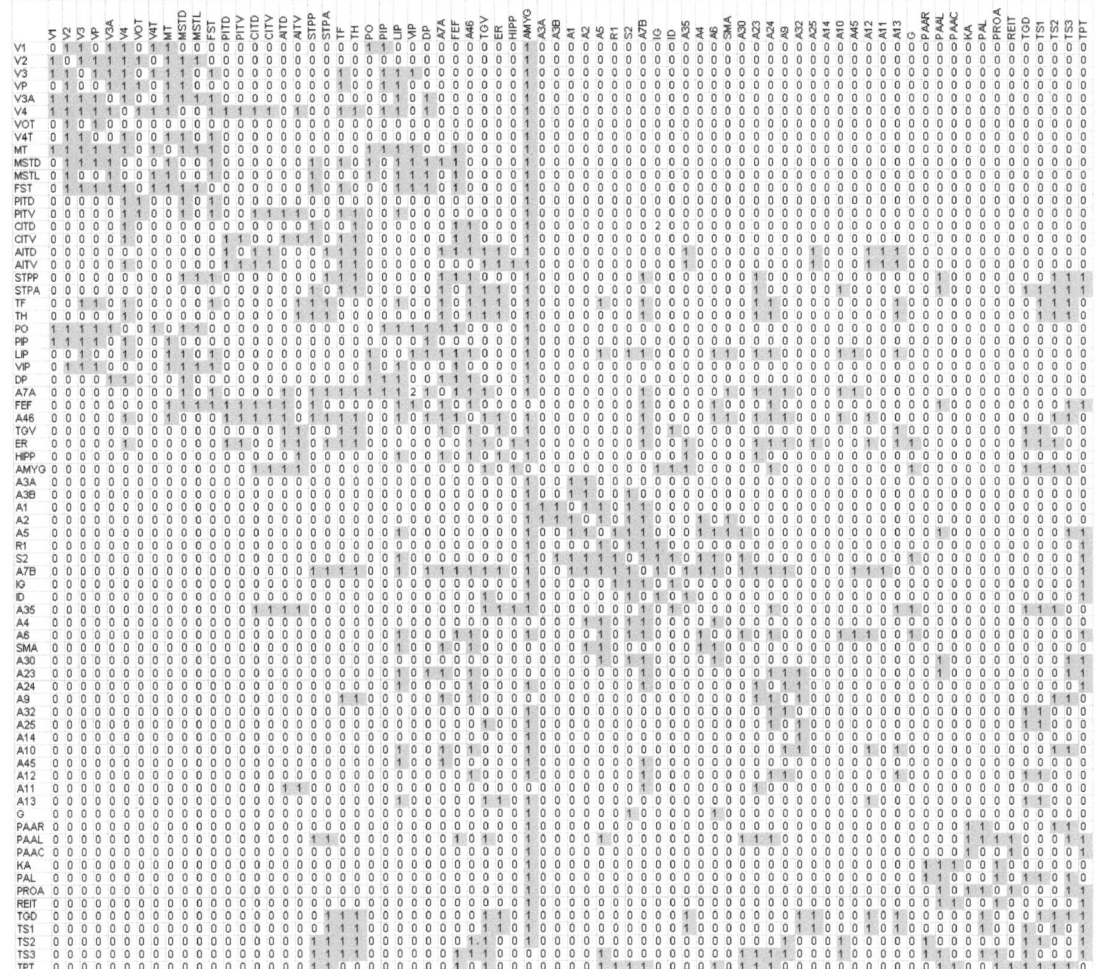

Figure A.3: Macaque connectivity matrix. The connectivity matrix consisted of 73 cortical regions and 837 projections (directed edges). A zero entry stands for connections that were either found absent or for which no search was done. Existing connections are assigned the value one in the adjacency matrix (and shown with gray background for easier visibility). Rows represent outgoing and columns incoming connections of a region.

Abbreviations

General Abbreviations

AS	Autonomous System
ASP	Average shortest path
CA1	Cornu ammonis (hippocampus) region 1
CA3	Cornu ammonis (hippocampus) region 3
DD	absolute difference of degrees
DTI	Diffusion Tensor Imaging
EEG	Electroencephalogram
EF	Edge frequency
fMRI	functional magnetic resonance imaging
MI	Matching index
MSA	Multi-perturbation Shapley value analysis
NMDS	Nonmetric multi-dimensional scaling
PD	Product of degrees
PET	Positron emission tomography

Macaque cortical areas

(Young et al., 1995; Stephan et al., 2000)

1	Primary somatosensory area 1
10	Area 10
11	Area 11
12	Area 12
13	Area 13
14	Area 14

ABBREVIATIONS

2	Primary somatosensory area 2
23	Cingulate area 23
24	Cingulate area 24
25	Area 25
32	Area 32
35	Perirhinal area 35
36	Area 36
3a	Primary somatosensory area 3a
3b	Primary somatosensory area 3b
4	Primary motor cortex, area 4
45	Prefrontal area 45
46	Prefrontal area 46
5	Somatosensory area 5
6	Premotor cortex, area 6
7a	Area 7a
7b	Area 7b
8a	Area 8a
8b	Area 8b
9	Prefrontal area 9
A	Periamygdalar allocortex
AITd	Anterior inferotemporal area (dorsal)
AITv	Anterior inferotemporal area (ventral)
AMYG	Amygdala
CITd	Central inferotemporal area (dorsal)
CITv	Central inferotemporal area (ventral)
DP	Dorsal prelunate area
ER	Entorhinal cortex
FA	Primary motor area FA
FB	Dorsolateral and medial premotor cortex
FBA	Ventrolateral premotor cortex (posterior part)
FCBm	Ventrolateral premotor cortex (anterior part)

ABBREVIATIONS

FCop	Frontal operculum
FEF	Frontal eye fields, area 8
FL	Subcallosal area FL
FST	Floor of superior temporal cortex
G	Gustatory area
HIPP	Hippocompus
IA	Insular area IA
IB	Insular area IB
Id	Dysgranular insula
Ig	Granular insula area
KA	Auditory koniocortical area
LA	Cingulate area LA
LC	Cingulate area LC
LIP	Lateral intraparietal area
MDP	Medial dorsal parietal area
MIP	Medial intraparietal area
MT	Middle temporal area, area V5
MSTd	Medial superior temporal area (dorsal)
MSTl	Medial superior temporal area (lateral)
OA	Prestriate occipital area OA, approximately area 19
OB	Prestriate occipital area OB, approximately area 18
OC	Striate occipital area OC, area 17, V1
PaAc	Auditory parakoniocortical area (caudal)
PaAl	Auditory parakoniocortical area (lateral)
PaAr	Auditory parakoniocortical area (rostral)
PB	Postcentral primary sensory area PB
PC	Primary sensory area PC
PCop	Parietal operculum
PEm	Anterior part of the superior parietal lobule
PEp	Posterior part of the superior parietal lobule
PF	Anterior part of the inferior parietal lobule

ABBREVIATIONS

PG	Posterior part of the inferior parietal lobule
PITd	Posterior inferotemporal area (dorsal)
PITv	Posterior inferotemporal area (ventral)
PIP	Posterior intraparietal area
PO	Parieto-occipital visual area
proA	Auditory prokoniocortical area
Reit	Auditory retroinsular temporal cortex
Ri	Retroinsular cortex
SII	Second somatosensory area
SMA	Supplementary motor area
STPa	Superior temporal polysensory area (anterior)
STPp	Superior temporal polysensory area (posterior)
TA	Temporal area TA
TB	Supratemporal area TB
TC	Supratemporal auditory koniocortical area TC
TE	Inferotemporal area TE
TEO	Inferotemporal area TEO
TF	Parahippocampal area TF
TG	Temporal pole area TG
TGd	Temporal polar cortex (dorsal)
TH	Parahippocampal area TH
TPt	Area TPt
TS1	Superior temporal auditory area 1
TS2	Superior temporal auditory area 2
TS3	Superior temporal auditory area 3
V1	Visual area 1, area 17
V2	Visual area 2
V3	Visual area 3
V3A	Visual area 3A
V4	Visual area 4
V4t	Visual area 4 transitional

VIP	Ventral intraparietal area
VOT	Ventral occipito-temporal area
VP	Ventral posterior area

Cat cortical areas

(Scannell et al., 1995; Scannell et al., 1999)

1	Area 1
17	Area 17
18	Area 18
19	Area 19
2	Area 2
20a	Area 20a
20b	Area 20b
21a	Area 21a
21b	Area 21b
35	Area 35
36	Area 36
3a	Area 3a
3b	Area 3b
4	Areas 4f, 4sf, 4d
4g	Area 4g
5m	Medial area 5
5Al	Lateral area 5A
5Am	Medial area 5A
5Bl	Lateral area 5B
5Bm	Medial area 5B
6l	Lateral area 6
6m	Medial area 6
7	Area 7
AAF	Anterior auditory field
AES	Anterior ectosylvian sulcus

ABBREVIATIONS

AI	Primary auditory field
AII	Second auditory field
ALLS	Anterolateral suprasylvian area
AMLS	Anteromedial lateral suprasylvian area
Amyg	Amygdala
CGa	Anterior cingulate cortex
CGp	Posterior cingulate cortex
DLS	Dorsolateral suprasylvian area
Enr	Entorhinal cortex
EPp	Posterior part of posterior ectosylvian cortex
Hipp	Hippocampus proper
IA	Agranular insula
IG	Granular insula
P	Posterior auditory field
PFCl	Lateral prefrontal cortex
PFCMd	Dorsal medial prefrontal cortex
PFCMil	Infralimbic medial prefrontal cortex
PLLS	Posterolateral suprasylvian area
PMLS	Posteromedial lateral suprasylvian area
PS	Posterior suprasylvian area
pSb	Presubiculum, parasubiculum, postsubicular cortex
RS	Retrosplenial cortex
Sb	Subiculum
SII	Second somatosensory area
SIV	Forth somatosensory area
SSAi	Inner (deep) suprasylvian sulcal region of area 5
SSAo	Outer suprasylvian sulcal region of area 5
SVA	Splenial visual area
Tem	Temporal auditory field
VLS	Ventrolateral suprasylvian area
VP(ctx)	Ventroposterior auditory field

Bibliography

Albert, R. & Barabási, A.-L. (2002). Statistical mechanics of complex networks. *Rev. Mod. Phys.*, 74(1):47–97.

Albert, R., Jeong, H., & Barabási, A.-L. (1999). Diameter of the World-Wide Web. *Nature*, 401:130–131.

Amaral, L. A. N., Scala, A., Barthélémy, M., & Stanley, H. E. (2000). Classes of small-world networks. *Proc. Natl. Acad. Sci.*, 97(21):11149–11152.

Andras, P., Panzeri, S., & Young, M. P. (2002). Towards statistically valid population decoding models. *Neurocomputing*, 44–46:269–274.

Barabási, A.-L. (2002). *Linked*. Perseus Publishing.

Barabási, A.-L. & Albert, R. (1999). Emergence of scaling in random networks. *Science*, 286:509–512.

Barabási, A.-L., Ravasz, E., & Vicsek, T. (2001). Deterministic scale-free networks. *Physica A*, 3–4:559–564.

Bartels, A. & Zeki, S. (2005). Brain dynamics during natural viewing conditions— a new guide for mapping connectivity in vivo. *Neuroimage*, 24:339–349.

Binzegger, T., Douglas, R. J., , & Martin, K. A. C. (2004). A quantitative map of the circuit of cat primary visual cortex. *J. Neurosci.*, 24:8441–8453.

Braitenberg, V. & Schüz, A. (1998). *Cortex: Statistics and Geometry of Neuronal Connectivity*. Springer, 2nd edition.

Brodmann, K. (1909). *Vergleichende Lokalisationslehre der Grosshirnrinde in ihren Prinzipien dargestellt auf Grund des Zellenbaues.* Barth, Leipzig.

Büchel, C. & Friston, K. J. (1997). Modulation of connectivity in visual pathways by attention: Cortical interactions evaluated with structural equation modelling and fMRI. *Cereb. Cortex.*, 7:768–778.

Burda, Z., Correia, J. D., & Krzywicki, A. (2001). Statistical ensemble of scale-free random graphs. *Phys. Rev. E*, 64:046118.

Caldarelli, G., Capocci, A., De Los Rios, P., & Munoz, M. A. (2002). Scale-free networks from varying vertex intrinsic fitness. *Phys. Rev. Lett.*, 89(25):258702.

Carmichael, S. T. & Price, J. L. (1994). Architectonic subdivision of the orbital and medial prefrontal cortex in the macaque monkey. *J. Comp. Neurol.*, 346:366–402.

Cherniak, C. (1994). Component placement optimization in the brain. *J. Neurosci.*, 14(4):2418–2427.

Cherniak, C. (1995). Neural component placement. *Trends Neurosci.*, 18:522–527.

Cherniak, C., Changizi, M., & Kang, D. W. (1999). Large-scale optimization of neuron arbors. *Phys. Rev. E*, 59:6001–6009.

Chklovskii, D. B., Schikorski, T., & Stevens, C. F. (2002). Wiring optimization in cortical circuits. *Neuron*, 34:341–347.

Choe, Y., McCormcik, B. H., & Koh, W. (2004). Network connectivity analysis on the temporally augmented *C. elegans* web: A pilot study. *Soc. Neurosci. Abstr.*, 30:921.9.

Coover, G. D., Murison, R., & Jellestad, F. K. (1992). Subtotal lesions of the amygdala: The rostral central nucleus in passive avoidance and ulceration. *Physiol. Behav.*, 51:795–803.

Cormen, T. H., Leiserson, C. E., Rivest, R. L., & Stein, C. (2001). *Introduction to Algorithms.* MIT Press, Cambridge, 2nd edition.

Damier, P., Hirsch, E. C., Agid, Y., & Graybiel, A. M. (1999). The substantia nigra of the human brain. II. patterns of loss of dopamine-containing neurons in parkinson's disease. *Brain*, 122:1437–1448.

Dayan, P. & Abbott, L. F. (2001). *Theoretical Neuroscience: Computational and Mathematical Modeling of Neural Systems.* MIT Press, Cambridge.

de Aguiar, M. A. M. & Bar-Yam, Y. (2005). Spectral analysis and the dynamic response of complex networks. *Phys. Rev. E*, 71:016106.

Diestel, R. (1997). *Graph Theory.* Springer, New York.

Durbin, R. M. (1987). *Studies on the Development and Organization of the Nervous System of* Caenorhabditis elegans. PhD thesis, University of Cambridge.

Ebel, H., Mielsch, L.-I., & Bornholdt, S. (2002). Scale-free topology of e-mail networks. *Phys. Rev. E*, 66:035103.

Eguíluz, V. M., Chialvo, D. R., Cecchi, G., Baliki, M., & Apkarian, A. V. (2005). Scale-free brain functional networks. *Phys Rev Lett*, 94:018102.

Erdös, P. & Rényi, A. (1960). On the evolution of random graphs. *Publ. Math. Inst. Hung. Acad. Sci.*, 5:17–61.

Felleman, D. J. & van Essen, D. C. (1991). Distributed hierarchical processing in the primate cerebral cortex. *Cereb. Cortex*, 1:1–47.

Finger, S. & Stein, D. G. (1982). *Brain Damage and Recovery.* Academic Press.

Friston, K. J. (1994). Functional and effective connectivity in neuroimaging: A synthesis. *Hum. Brain Mapp.*, 2:56–78.

Gavin *et al.* (2002). Functional organization of the yeast proteome by systematic analysis of protein complexes. *Nature*, 415:141–147.

Geschwind, N. (1965). Disconnection syndromes in animals and man: Part I. *Brain*, 88:229–237.

Girvan, M. & Newman, M. E. J. (2002). Community structure in social and biological networks. *Proc. Natl. Acad. Sci.*, 99(12):7821–7826.

Goldberg, D. S. & Roth, F. P. (2003). Assessing experimentally derived interactions in a small world. *Proc. Natl. Acad. Sci. USA*, 100:4372–4376.

Granovetter, M. S. (1973). The strength of weak ties. *Am. J. Sociol.*, 78(6):1360–1380.

Grigorov, M. G. (2005). Global properties of biological networks. *Drug Discovery Today*, 10:365–372.

Gross, J. & Yellen, J. (1998). *Graph Theory and its Applications*. CRC Press.

Han, J.-D. J., Bertin, N., Hao, T., Goldberg, D. S., Berriz, G. F., Zhang, L. V., Dupuy, D., Walhout, A. J. M., Cusick, M. E., Roth, F. P., & Vidal, M. (2004). Evidence for dynamically organized modularity in the yeast protein-protein interaction network. *Nature*, 430:88–93.

Hellwig, B. (2000). A quantitative analysis of the local connectivity between pyramidal neurons in layers 2/3 of the rat visual cortex. *Biol. Cybern.*, 82:111–121.

Hilgetag, C. C. (2000). Spatial neglect and paradoxical lesion effects in the cat — a model based on midbrain connectivity. *Neurocomputing*, 32–33:793–799.

Hilgetag, C. C., Burns, G. A. P. C., O'Neill, M. A., Scannell, J. W., & Young, M. P. (2000). Anatomical connectivity defines the organization of clusters of cortical areas in the macaque monkey and the cat. *Phil. Trans. R. Soc. Lond. B*, 355:91–110.

Hilgetag, C. C. & Kaiser, M. (2004). Clustered organisation of cortical connectivity. *Neuroinformatics*, 2:353–360.

Hilgetag, C. C., Kötter, R., Stephan, K. E., & Sporns, O. (2002). Computational methods for the analysis of brain connectivity. In *Computational Neuroanatomy*, Chapter 14, pages 295–335. Humana Press, Totowa, NJ.

Hinrichsen, K. (1990). *Humanembryologie*. Springer, Berlin.

His, W. (1888). Zur Geschichte des Gehirns sowie der centralen und peripherischen Nervenbahnen beim menschlichen Embryo. *Abhandlungen der mathematisch-physikalischen Classe der Königlichl. Sächsichen Gesellschaft der Wissenschaften*, 14(7).

Ho *et al.* (2002). Systematic identification of protein complexes in *Saccharomyces cerevisae* by mass spectrometry. *Nature*, 415:180–183.

Holme, P., Kim, B. J., Yoon, C. N., & Han, S. K. (2002). Attack vulnerability of complex networks. *Phys. Rev. E*, 65:056109.

Holme, P. & Kimy, B. J. (2002). Growing scale-free networks with tunable clustering. *Phys. Rev. E*, 65(2):026107.

Hubel, D., Wiesel, T., & LeVay, S. (1977). Plasticity of ocular dominance columns in monkey striate cortex. *Philos. Trans. R. Soc. Lond. Ser. B*, 278:377–409.

Huberman, B. A. & Adamic, L. A. (1999). Growth dynamics of the world-wide web. *Nature*, 401:131.

Ito, T., Chiba, T., Ozawa, R., Yoshida, M., Hattori, M., , & Sakaki, Y. (2001). A comprehensive two-hybrid analysis to explore the yeast protein interactome. *Proc. Natl. Acad. Sci.*, 98(8):4569–4574.

Izhikevich, E. M. (2003). Simple model of spiking neurons. *IEEE Trans. Neural Networks*, 14:1569–1572.

Izhikevich, E. M. (2004). Which model to use for cortical spiking neurons? *IEEE Trans. Neural Networks*, 15:1063–1070.

Jaeger, H. & Haas, H. (2004). Harnessing nonlinearity: Predicting chaotic systems and saving energy in wireless communication. *Science*, 304:78–80.

Jeong, H., Mason, S. P., Barabási, A.-L., & Oltvai, Z. N. (2001). Lethality and centrality in protein networks. *Nature*, 411:41–42.

Jeong, H., Tombor, B., Albert, R., Oltwal, Z., & Barabási, A.-L. (2000). The large-scale organization of metabolic networks. *Nature*, 407:651–654.

Kahle, W. (1969). *Die Entwicklung der Menschlichen Großhirnhemisphäre*. Springer, Berlin.

Kaiser, M. & Hilgetag, C. C. (2004a). Edge vulnerability in neural and metabolic networks. *Biol. Cybern.*, 90:311–317.

Kaiser, M. & Hilgetag, C. C. (2004b). Modelling the development of cortical networks. *Neurocomputing*, 58–60:297–302.

Kaiser, M. & Hilgetag, C. C. (2004c). Spatial growth of real-world networks. *Phys. Rev. E*, 69:036103.

Kaiser, M., Martin, R., Andras, P., & Young, M. P. (2005). Structural robustness of cortical networks. *J. Neurosci.*, page (submitted).

Kandel, E. R., Schwartz, J. H., & Jessell, T. M., Editoren (2000). *Principles of Neural Science*. McGraw-Hill, 4th edition.

Keinan, A., Sandbank, B., Hilgetag, C. C., Meilijson, I., & Ruppin, E. (2004). Fair attribution of functional contribution in artificial and biological networks. *Neural Comp.*, 16:1887–1915.

Kitano, H. (2002). Computational systems biology. *Nature*, 420:206–210.

Klyachko, V. A. & Stevens, C. F. (2003). Connectivity optimization and the positioning of cortical areas. *Proc. Natl. Acad. Sci. USA*, 100:7937–7941.

Kötter, R. (2004). Online retrieval, processing, and visualization of primate connectivity data from the CoCoMac database. *Neuroinformatics*, 2:127–144.

Kötter, R. & Stephan, K. E. (2003). Network participation indices: Characterizing component roles for information processing in neural networks. *Neural Networks*, 16:1261–1275.

Kozloski, J., Hamzei-Sichani, F., & Yuste, R. (2001). Stereotyped position of local synaptic targets in neocortex. *Science*, 293:868–872.

Krubitzer, L. & Kahn, D. M. (2003). Nature versus nurture revisited: An old idea with a new twist. *Prog. Neurobiol.*, 70:33–52.

Lewis, J. & Van Essen, D. (2000). Architectonic parcellation of parieto-occipital cortex and interconnected cortical regions in the macaque monkey. *J. Comp. Neurol.*, 428:79–111.

Lieberman, E., Hauert, C., & Nowak, M. A. (2005). Evolutionary dynamics on graphs. *Nature*, 433:312–316.

Liljeros, F., Edling, C. R., Amaral, L. A. N., Stanley, H. E., & Åberg, Y. (2001). The web of human sexual contacts. *Nature*, 411:907–908.

Milgram, S. (1967). The small-world problem. *Psychology Today*, 1:60–67.

Muller, J., Corodimas, K. P., Fridel, Z., & LeDoux, J. E. (1997). Functional inactivation of the lateral and basal nuclei of the amygdala by muscimol infusion prevents fear conditioning to an explicit conditioned stimulus and to contextual stimuli. *Behav. Neurosci.*, 111:683–691.

Murray, J. D. (1990). *Mathematical Biology*. Springer, Heidelberg.

Newman, M. E. J. (2003). The structure and function of complex networks. *SIAM Review*, 45(2):167–256.

Panzeri, S., Schultz, S., Treves, A., & Rolls, E. (1999). Correlations and the encoding of information in the nervous system. *Proc. R. Soc. Lond. Ser. B*, 266:1001–1012.

Papp, B., Pál, C., & Hurst, L. D. (2004). Metabolic network analysis of the causes and evolution of enzyme dispensability in yeast. *Nature*, 429:661–664.

Parker, G. J. M., Stephan, K. E., Barker, G. J., Rowe, J. B., MacManus, D. G., Wheeler-Kingshott, C. A. M., Ciccarelli, O., Passingham, R. E., Spinks, R. L., Lemon, R. N., & Turner, R. (2002). Initial demonstration of *in vivo* tracing of axonal projections in the macaque brain and comparison with the human brain using diffusion tensor imaging and fast marching tractography. *Neuroimage*, 15:797–809.

Passingham, R. E., Stephan, K. E., & Kötter, R. (2002). The anatomical basis of functional localization in the cortex. *Nat. Rev. Neurosci.*, 3:606–616.

Patterson, D., Brown, A., Broadwell, P., Candea, G., Chen, M., Cutler, J., Enriquez, P., Fox, A., Kýcýman, E., Merzbacher, M., Oppenheimer, D., Sastry, N., Tetzlaff, W., Traupman, J., & Treuhaft, N. (2002). Recovery Oriented Computing (ROC): Motivation, Definition, Techniques, and Case Studies. Technical report, Computer Science Technical Report UCB//CSD-02-1175, U.C. Berkeley.

Petroni, F., Panzeri, S., Hilgetag, C. C., Kötter, R., & Young, M. P. (2001). Simultaneity of responses in a hierarchical visual network. *Neuroreport*, 12:2753–2759.

Preuss, T. (2000). What's human about the human brain. In Gazzaniga, M., Editor, *The New Cognitive Neurosciences*, pages 1219–1234. MIT Press, Cambridge, MA.

Rakic, P. (2002). Neurogenesis in adult primate neocortex: An evaluation of the evidence. *Nature Rev. Neurosci.*, 3:65–71.

Ravasz, E., Somera, A. L., Mongru, D. A., Oltvai, Z. N., & Barabási, A.-L. (2002). Hierarchical organization of modularity in metabolic networks. *Science*, 297:1551–1555.

Reigl, M., Alon, U., & Chklovskii, D. B. (2004). Search for computational modules in the *C. elegans* brain. *BMC Biology*, 2:/1741-7007-2-25.

Scannell, J., Blakemore, C., & Young, M. (1995). Analysis of connectivity in the cat cerebral cortex. *J. Neurosci.*, 15(2):1463–1483.

Scannell, J. W., Burns, G. A., Hilgetag, C. C., O'Neil, M. A., & Young, M. P. (1999). The connectional organization of the cortico-thalamic system of the cat. *Cereb. Cortex*, 9(3):277–299.

Schuster, S. & Hilgetag, C. (1994). On elementary flux modes in biochemical reaction systems at steady state. *J. Biol. Systems*, 2:165–182.

Schuster, S., Hilgetag, C., Woods, J. H., & Fell, D. A. (2002). Reaction routes in biochemical reaction systems: Algebraic properties, validated calculation procedure and example from nucleotide metabolism. *J. Math. Biol.*, 45:153–181.

Schwikowski, B., Uetz, P., & Fields, S. (2000). A network of protein-protein interactions in yeast. *Nature Biotech.*, 18:1257–1261.

Seth, A. K. & Edelman, G. M. (2004). Environment and behavior influence the complexity of evolved neural networks. *Adaptive Behavior*, 12:5–20.

Shannon, C. E. (1948). A mathematical theory of communication. *The Bell Systems Technical Journal*, 27:379–423, 623–656.

Shon, J., Park, J. Y., & Wei, L. (2003). Beyond similarity-based methods to associate genes for the inference of function. *BIOSILICIO*, 1(3):89–96.

Singer, W. (1993). Synchronization of cortical activity and its putative role in information processing and learning. *Annu. Rev. Physiol.*, pages 349–374.

Smith, V., Lashkari, K. C. D., Botstein, D., & Brown, P. (1997). Functional analysis of the genes of yeast chromosome V by genetic footprinting. *Science*, 275:2069–2074.

Song, S., Sjöström, P. J., Reigl, M., Nelson, S., & Chklovskii, D. B. (2005). Highly nonrandom features of synaptic connectivity in local cortical circuits. *PLoS Biol.*, 3:e68.

Spear, P., Tong, L., & McCall, M. (1988). Functional influence of areas 17, 18 and 19 on lateral suprasylvian cortex in kittens and adult cats: implications for compensation following early visual cortex damage. *Brain Res.*, 447(1):79–91.

Sperry, R. W. (1963). Chemoaffinity in the orderly growth of nerve fiber pattern and connections. *Proc. Natl. Acad. Sci. USA*, 50:703–710.

Sporns, O. (2002). Graph theory methods for the analysis of neural connectivity patterns. In *Neuroscience Databases*, Chapter 12, pages 169–183. Kluwer Academic, Dordrecht.

Sporns, O., Chialvo, D. R., Kaiser, M., & Hilgetag, C. C. (2004). Organization, development and function of complex brain networks. *Trends Cogn. Sci.*, 8:418–425.

Sporns, O. & Kötter, R. (2004). Motifs in brain networks. *PLOS Biology*, 2:1910–1918.

Sporns, O., Tononi, G., & Edelman, G. M. (2000a). Connectivity and complexity: The relationship between neuroanatomy and brain dynamics. *Neural Networks*, 13:909–922.

Sporns, O., Tononi, G., & Edelman, G. M. (2000b). Theoretical neuroanatomy: Relating anatomical and functional connectivity in graphs and cortical connection matrices. *Cereb. Cortex*, 10:127–141.

Sporns, O. & Zwi, J. D. (2004). The small world of the cerebral cortex. *Neuroinformatics*, 2:145–162.

Stelling, J., Klamt, S., Bettenbrock, K., Schuster, S., & Gilles, E. D. (2002). Metabolic network structure determines key aspects of functionality and regulation. *Nature*, 420:190–193.

Stephan, K. E., Hilgetag, C. C., Burns, G. A. P. C., O'Neill, M. A., Young, M. P., & Kötter, R. (2000). Computational analysis of functional connectivity between areas of primate cerebral cortex. *Phil. Trans. R. Soc.*, 355:111–126.

Strogatz, S. H. (2001). Exploring complex networks. *Nature*, 410:268–276.

Stromswold, K. (2000). The cognitive neuroscience of language acquisition. In Gazzaniga, M., Editor, *The New Cognitive Neurosciences*, pages 909–932. MIT Press, Cambridge, MA, 2nd edition.

Sur, M. & Leamey, C. A. (2001). Development and plasticity of cortical areas and networks. *Nature Rev. Neurosci.*, 2:251–262.

Szabó, P., Scheuring, I., Czárán, T., & Szathmáry, E. (2002). *In silico* simulations reveal that replicators with limited dispersal evolve towards higher efficiency and fidelity. *Nature*, 420:340–343.

Tanenbaum, A. S. (2003). *Computer Networks*. PH PTR, 4th edition.

Thompson, D. (2004). *On Growth and Form*. Cambridge University Press.

Trappenberg, T. P. (2002). *Fundamentals of Computational Neuroscience*. Oxford University Press.

Tuch, D. S., Reese, T. G., Wiegell, M. R., & Wedeen, V. J. (2003). Diffusion MRI of complex neural architecture. *Neuron*, 40:885–895.

Uetz, P., Giot, L., Cagney, G., Mansfield, T. A., Judson, R. S., Knight, J. R., Lockshon, D., Narayan, V., Srinivasan, M., Pochart, P., Qureshi-Emili, A., Li, Y., Godwin, B., Conover, D., Kalbfleisch, T., Vijayadamodar, G., Yang, M., Johnston, M., Fields, S., & Rothberg, J. M. (2000). A comprehensive analysis of protein-protein interactions in *Saccharomyces cerevisiae* . *Nature*, 403:623–627.

Ungerleider, L. & Mischkin, M. (1982). Two cortical visual systems. In Ingle, M., Goodale, M., & Mansfield, R., Editoren, *The New Cognitive Neurosciences*. MIT Press, Cambridge, MA.

Valverde, S., Cancho, R. F., & Solé, R. V. (2002). Scale-free networks from optimal design. *Europhys. Lett.*, 60(4):512–517.

Verhage, M., Maia, A. S., Plomp, J. J., Brussaard, A. B., Heeroma, J. H., Vermeer, H., Toonen, R. F., Hammer, R. E., van den Berg, T. K., Missler, M., Geuze, H. J., & Südhof, T. C. (2000). Synaptic assembly of the brain in the absence of neurotransmitter secretion. *Science*, 287:864–869.

von der Malsburg, C. (1981). The correlation theory of brain function. Technical report, Max-Planck-Institute for Biophysical Chemistry.

von Neumann, J. (1958). *The Computer and the Brain*. Yale University Press.

Wagner, A. (2000). Robustness against mutations in genetic networks of yeast. *Nature Genetics*, 24:355–361.

Wagner, A. (2005). Robustness, evolvability, and neutrality. *FEBS Lett.*, 579:1772–1778.

Watts, D. J. (1999). *Small Worlds*. Princeton University Press, Princeton.

Watts, D. J., Dodds, P. S., & Newman, M. E. J. (2002). Identity and search in social networks. *Science*, 296:1302–1305.

Watts, D. J. & Strogatz, S. H. (1998). Collective dynamics of 'small-world' networks. *Nature*, 393:440–442.

Waxman, B. M. (1988). Routing of multipoint connections. *IEEE J. Sel. Areas Commun.*, 6(9):1617–1622.

Weiss, P. (1941). Nerve patterns: The mechanisms of nerve growth. *Growth*, 5(Suppl):163–203.

White, J. G., Southgate, E., Thomson, J. N., & Brenner, S. (1986). The structure of the nervous system of the nematode *Caenorhabditis elegans*. *Phil. Trans. R. Soc. London Ser. B*, 314(1165):1–340.

Wiener, N. (1961). *Cybernetics: Or Control and Communication in the Animal and the Machine*. MIT Press.

Yook, S.-H., Jeong, H., & Barabási, A.-L. (2002). Modeling the Internet's large-scale topology. *Proc. Natl. Acad. Sci.*, 99(21):13382–13386.

You, S. W., Chen, B., Liu, H., Lang, B., Xia, J., Jiao, X., & Ju, G. (2003). Spontaneous recovery of locomotion induced by remaining fibers after spinal cord transection in adult rats. *Restor. Neurol. Neurosci.*, 21:39–45.

Young, M. P. (1992). Objective analysis of the topological organization of the primate cortical visual system. *Nature*, 358(6382):152–155.

Young, M. P. (1993). The organization of neural systems in the primate cerebral cortex. *Phil. Trans. R. Soc.*, 252:13–18.

Young, M. P. (2000). The architecture of visual cortex and inferential processes in vision. *Spatial Vision*, 13(2–3):137–146.

Young, M. P., Hilgetag, C. C., & Scannell, J. W. (1999). Models of paradoxical lesion effects and rules of inference for imputing function to structure in the brain. *Neurocomputing*, 26–27:933–938.

Young, M. P., Hilgetag, C. C., & Scannell, J. W. (2000). On imputing function to structure from the behavioural effects of brain lesions. *Phil. Trans. R. Soc.*, 355:147–161.

Young, M. P. & Scannell, J. W. (1996). Component-placement optimization in the brain. *Trends Neurosci.*, 19:413–415.

Young, M. P., Scannell, J. W., O'Neill, M. A., Hilgetag, C. C., Burns, G., & Blakemore, C. (1995). Non-metric multidimensional scaling in the analysis of neuroanatomical connection data and the organization of the primate cortical visual system. *Phil. Trans. R. Soc.*, 348:281–308.

Zilles, K. & Rehkämper, G. (1998). *Funktionelle Neuroanatomie*. Springer.

Zola-Morgan, S., Squire, L. R., & Amaral, D. G. (1986). Human amnesia and the medial temporal region: Enduring memory impairment following a bilateral lesion limited to field CA1 of the hippocampus. *J. Neurosci.*, 6:2950–2067.

Publication list

Peer-reviewed Journals

1. Kaiser M, Martin R, Andras P, Young MP. Structural robustness of cortical networks (submitted)

2. Kaiser M, Hilgetag CC. Test of optimal component placement in Macaque and *C. elegans* neural networks. (submitted)

3. Sporns O, Chialvo DR, Kaiser M, Hilgetag CC (2004) Organization, Development and Function of Complex Brain Networks. *Trends in Cognitive Sciences* 8:418-425

4. Hilgetag CC, Kaiser M (2004) Clustered organization of cortical connectivity. *Neuroinformatics* 2:353-360.

5. Kaiser M, Hilgetag CC (2004) Modelling the development of cortical networks. *Neurocomputing* 58-60:297-302.

6. Kaiser M, Hilgetag CC (2004) Edge vulnerability in cortical and biochemical networks. *Biological Cybernetics* 90:311-317.

7. Kaiser M, Hilgetag CC (2004) Spatial growth of real-world networks. *Physical Review E* 69:036103

8. Kaiser M, Lappe M (2004) Perisaccadic mislocalization orthogonal to saccade direction. *Neuron* 41:293-300.

Conference Abstracts

1. Kaiser M (2005) Which constraints shape brain connectivity? *9th International Conference on Cognitive and Neural Systems*

2. Kaiser M, Hilgetag CC (2005) Development and Robustness of Cortical Brain Networks. *69th Conference of the German Physics Society (DPG)*, SYBN 3.27

3. Hilgetag CC, Kaiser M (2005) Organization of Cortical Brain Networks. *69th Conference of the German Physics Society (DPG)*, SYBN 3.28

4. Kaiser M, Hilgetag CC (2005) Test of optimal component placement in Macaque and *C. elegans* neural networks. Conference of the *Computational Neuroscience Society*

5. Kaiser M, Hilgetag CC (2004) Clustered cortical architecture: Development and robustness. *Society for Neuroscience Abstracts*, Vol. 30

6. Kaiser M (2004) Structural patterns underlying robustness in neural systems. *8th International Conference on Cognitive and Neural Systems*

7. Kaiser M, Hilgetag CC (2004) Spatial development of cortical networks: Essential factors for generating clusters and long-range connections. *4th Forum of European Neuroscience (FENS)*

8. Oerke B, Kaiser M, Lappe M (2004) Seeing green on green. *7. Tübinger Wahrnehmungskonferenz (TWK)*

9. Kaiser M, Hilgetag CC (2003) Modelling the development of cortical networks. Conference of the *Computational Neuroscience Society*

10. Kaiser M, Hilgetag CC (2003) A model for the development of global cortico-cortical connectivity. *7th International Conference on Cognitive and Neural Systems*

11. Lappe M, Kaiser M, Awater H (2002) Visual Factors in Peri-saccadic Compression of Space. *Visual Localization in Space-Time* (pre-ECVP conference).

12. Kaiser M, Lappe M (2002) Perisaccadic compression of space orthogonal to saccade direction. *Vision Sciences Society Annual Conference* (Abstract published in Journal of Vision)

13. Kaiser M, Lappe M (2002) Perisakkadische Kompression in zwei Dimensionen. *5. Tübinger Wahrnehmungskonferenz (TWK)*

14. Martin R, Kaiser M, Andras P, Young MP (2001) Is the brain a scale-free network? *Society for Neuroscience Abstracts*, Vol. 27

Other publications

1. Hilgetag CC, Kaiser M (2005) Die Netzwerk-Struktur biologischer Systeme. *BIOforum* 04/2005:32–33

2. Kaiser M (2005) Spatial network growth: Generating small-world, scale-free, and multi-cluster spatial networks. *IUB School of Engineering and Science Technical Report* No. 1

Curriculum vitae

Name: Marcus Kaiser

Date of Birth: 20/January/1977 in Essen, Germany

1987 – 1996: Carl-Humann-Gymnasium (high school), Essen-Steele, Germany

1996 – 1997: Community service (Zivildienst)

1997 – 2002: Studying Biology, Ruhr-University Bochum, Germany

                MSc-Thesis: Computer simulation of visual localization during saccadic eye movements

                Supervisor: Prof. Dr. Markus Lappe

                Department of Zoology and Neurobiology headed by Prof. Dr. K.-P. Hoffmann

since 1998: Studying Computer science,

                FernUniversität Hagen (distance university), Germany

                leading to degree Master of Computer Science

2002 – 2005: PhD studies in Neuroscience,

                International University Bremen, Germany

                School of Engineering and Science

                Thesis: Neural and Biochemical Networks: Organization, Development, and Robustness

                Supervisor: Prof. Dr. Claus C. Hilgetag